FLORAL MIMICRY

T0177910

FLORAL MIMICRY

Steven D. Johnson
School of Life Sciences, University of KwaZulu-Natal, South Africa

Florian P. Schiestl
Department of Systematic and Evolutionary Botany, University of Zürich, Switzerland

OXFORD
UNIVERSITY PRESS

Great Clarendon Street, Oxford, OX2 6DP,
United Kingdom

Oxford University Press is a department of the University of Oxford.
It furthers the University's objective of excellence in research, scholarship,
and education by publishing worldwide. Oxford is a registered trade mark of
Oxford University Press in the UK and in certain other countries

© Steven D. Johnson & Florian P. Schiestl 2016

The moral rights of the authors have been asserted

First Edition published in 2016
Impression: 1

All rights reserved. No part of this publication may be reproduced, stored in
a retrieval system, or transmitted, in any form or by any means, without the
prior permission in writing of Oxford University Press, or as expressly permitted
by law, by licence or under terms agreed with the appropriate reprographics
rights organization. Enquiries concerning reproduction outside the scope of the
above should be sent to the Rights Department, Oxford University Press, at the
address above

You must not circulate this work in any other form
and you must impose this same condition on any acquirer

Published in the United States of America by Oxford University Press
198 Madison Avenue, New York, NY 10016, United States of America

British Library Cataloguing in Publication Data
Data available

Library of Congress Control Number: 2016943650

ISBN 978–0–19–873269–3 (hbk.)
ISBN 978–0–19–873270–9 (pbk.)

DOI 10.1093/acprof:oso/9780198732693.001.0001

Printed and bound by
CPI Litho (UK) Ltd, Croydon, CR0 4YY

Dedicated to the memory of Stefan Vogel (1925–2015)

Preface

The initial outline of this book was sketched on a piece of paper while we were traveling in a bus in southern Italy in 2013. We were returning from a conference, and as the bus hurtled down yet another mountain pass with half-completed roadworks and precipitous drops below the hairpin bends, our ideas started to gain a different kind of momentum. John Thompson was sitting across the aisle from us and readily shared some of the practical experience he had gained from writing his well-known books on coevolution. His immediate enthusiasm for the concept of this book served as an important catalyst for the project at that nascent stage. By the time we arrived at the airport, the basic structure of the book had been formulated. But that was the easy part. . .

The hardest aspect of writing a scientific book is dealing with the ongoing explosion in knowledge. We quickly realized that it would be impossible to be encyclopedic in our coverage of the topic of mimicry—which is why there are very few summary tables in this book—and that it would be more useful to ourselves and to our readers if we could identify key concepts and then evaluate published studies in terms of their contributions to these concepts. This approach also makes more sense in the internet age where the traditional encyclopedia has largely been replaced with search engines, yet the role of a book as a story or "long argument" (to borrow from Darwin) remains as important as ever. Another motivation for writing this book was that the field of mimicry has often been characterized by adaptive storytelling; we wanted to take a more critical approach that emphasizes the importance of experimental evidence.

Mimicry can compete with any crime novel or television program in terms of the intrigue of the plot and the subterfuge of the main actors. Who could conceive of a plant with a flower that traps its pollinators, tricks them into laying eggs, and then eventually starves them and their offspring to death? Yet such a plant, the long-tailed wild ginger, *Asarum caudatum*, is out there, waiting for its next victim, and it is only one of hundreds of plant species that similarly imprison their pollinators. And then there are the orchids that use chemical wizardry to trick male insects into copulating with their flowers. These species, and the plethora of other floral mimics, defy the conventional wisdom that pollination is the outcome of benign cooperative mutualisms between plants and animals.

Evolutionary biology addresses the "why" questions in nature, the story behind the story, as it were. The study of mimicry begins with discovering "how" species interact, along with the signals and senses and behavior that they use to do so, and then leads naturally to the search for ultimate causes. It is these questions, such as why many flowers lack rewards, why some pollination systems are so specialized, and why some lineages are so rich in species, that inspire us the most. This is a book about evolutionary concepts rather than the biology of particular species, but we have included many examples of some of the wonderful natural history that accompanies this subject. We thus hope that we can satisfy the intellectual curiosity of the reader as well as inspire a sense of awe and astonishment. In this regard, we are enormously grateful to the numerous photographers who willingly supplied images for this book. They are credited alongside each image.

This book has benefited from the inputs of dozens of people in the research community. We are thankful to the anonymous reviewers of the original book concept, as their comments helped to shape the content and style of the book. We also benefited enormously from the constructive and detailed reviews of the individual chapters that were provided by the following people: Jon Ågren, Bruce Anderson, Hans Bänziger, Judie Bronstein, Karl Duffy, Lawrence Harder, Nina Hobbhahn, Jana Jersáková, Cris Kuhlemeier, Adrian Leuchtmann, David Kikuchi, Timo van der Niet, Rod Peakall, Ryan Phillips, Rob Raguso, Graeme Ruxton, Philipp Schlüter, Adam Shuttleworth, and Johannes Späthe. Other people willingly provided advice on difficult concepts and in this regard we thank Stefan Dötterl, Allan Ellis, Craig Peter, Tobias Policha, Bitty Roy, Roger Seymour, and John Skelhorn. We are also very grateful to Kathryn Johnson for proof-reading the final text.

It has been a great pleasure to work with the team at Oxford University Press. Ian Sherman showed great enthusiasm for the project and Lucy Nash patiently guided the development of the book. Other staff will play a further role in production after the writing of this preface, and we thank them in advance.

The process of book-writing extends after hours and over weekends and is not much fun for the rest of the family. In this regard, we are deeply grateful to Kathryn, Daniel and Amy Johnson, and Simone Berchtold Schiestl and Luis and Moritz Schiestl for their patience, understanding, and support.

Finally, this book is dedicated to the memory of Stefan Vogel and we would like to remind readers of some the diverse contributions that he made to our field. Vogel, who died in 2015, was one of the greatest pollination biologists of the twentieth century. He was a phenomenally talented field biologist and scientific illustrator and made a large number of original discoveries, such as the oil bee pollination system and floral mimicry of fungi. Vogel also pioneered the functional understanding and analysis of floral fragrance using chromatographic tools. His monographs and journal articles will inspire research on the topic of plant–pollinator interactions for generations to come.

Contents

CHAPTER 1

Introduction

An encounter in a steamy rain forest with a flowering *Rhizanthes zippelii* makes an intruder wonder. The look of the flowers is more akin to a tentacled animal—a starfish or medusa—than to a member of the plant kingdom. But at the same time the reddish globe with crater, embedded in a tangle of rufous hairs, reminds one of the blood-shot orifice of a furry mammal. This perplexing aspect is part of an intricate set of lures: visual, tactile, olfactory, and gustatory. . . . . . . Behind this disconcerting look hide potent insect-manipulating powers to entice, appease and deceive a disparate cohort of nectar thieves, opportunists, female-chasing males and predators, as well as dupes which lay hundreds of ill-fated eggs on the flowers.

Bänziger (1995, p. 337; 1996a, p. 113)

What is mimicry?

The discovery by Henry Walter Bates (1862) of wing pattern mimicry among butterflies of the Amazon Basin was an instant sensation, providing Darwin and his followers with a favorite example of adaptation through natural selection. Even today, most biologists, when asked to give an example of mimicry, think first of palatable butterflies that mimic unpalatable ones, thereby obtaining increased protection from predators; few think first of a plant example. Yet mimicry is common among plants, not only flowering plants but also gymnosperms and mosses, and even occurs in other kingdoms such as the fungi. This book is about floral mimicry but also mimicry in general because, as we emphasize in this introductory chapter, all mimicry systems share common evolutionary principles.

In essence, mimicry is a phenomenon in nature whereby organisms use the signals of other organisms to mislead certain animals about their true identity. By deploying imitative signals, mimics are able to exploit the behavioral responses of those animals to the particular phenotypes of their models. These responses, which can include avoidance, attraction, probing, and even copulation or oviposition, ultimately enhance the fitness of the mimic. The link between mimetic traits and fitness is particularly obvious and is a major reason why mimicry has become such a classic example of the power of natural selection to shape the phenotypes of organisms.

The field of mimicry research has been riven with debates over definitions—so much so that Miriam Rothschild (1981) once referred to "mimicrats," a special category of biologists who focus on definitions of mimicry! While we are loathe to join their ranks, the reality is that many of the examples of floral mimicry covered in this book cannot be accommodated within current definitions. For example, a very useful and widely used definition of mimicry by the British lepidopterist Vane-Wright (1980) makes reference to models as living organisms, yet some of the well-documented floral mimicry systems involve models that are either dead (i.e., carrion) or the by-products of living organisms (e.g., feces). For the purposes of this book, therefore, we build on some excellent earlier attempts by Vane-Wright (1980) and others, and define mimicry as

The adaptive resemblance of one organism (the mimic) to other organisms or their by-products (the models),

Floral Mimicry. Steven D. Johnson & Florian P. Schiestl.
© Steven D. Johnson & Florian P. Schiestl 2016. Published 2016 by Oxford University Press.
DOI 10.1093/acprof:oso/9780198732693.001.0001

such that there is cognitive misclassification and behaviour by third-party organisms (operators) that enhances the fitness of the mimic.

This working definition incorporates the role of adaptation, and thus excludes coincidental resemblance among organisms. It excludes other types of mimicry in nature (such as biochemical mimicry by pathogens) that do not involve cognition by a sentient third party. It also excludes camouflage and crypsis, which do not involve detection and cognitive misclassification. It does allow for the concept of masquerade, a special case of mimicry involving an adaptive resemblance to an object such that it is detected and cognitively misclassified as being of no interest to an operator (Endler 1981; Skelhorn et al. 2010). In theory, masquerade can involve mimicry of inorganic objects and thus extend beyond this definition, but we are not aware of any convincing cases as yet. We specifically included biological by-products to account for mimicry of feces and carrion, objects of interest to certain animals searching for brood sites and food. Finally, we emphasize that the word "resemblance" in this definition applies to any signals, for example olfactory or auditory, not just visual ones.

The phenomenon of mimicry continues to grip the imagination of biologists and the general public alike. It plays into our own fears of being deceived—think of the Trojan horse, forgery of artworks, or a computer virus lurking within an innocuously named file. Above all, it serves as a vivid reminder of the cold, yet compelling, logic of evolution. The biologist and statistician Ronald Fisher (1930, p. 163) famously considered the theory of mimicry to be "the greatest post-Darwinian application of natural selection." Importantly, Fisher wrote this at a time when, as he put it, "the concentration of biological efforts in the museum and laboratories" had moved away from field-based natural history. He reminded his readers that mimicry illustrated the "importance of ecological observations for the interpretation of the material gathered in the great museums." His assertion that the study of evolution requires a combination of field- and laboratory-based studies remains as true today as it was then, and is a constant theme throughout this book.

The traditional focus of mimicry research has been on defense in animals, but there is now also a highly developed and rapidly growing body of research on floral mimicry in plants. There has, however, never been a major synthesis of the research on floral mimicry.

The scope of floral mimicry

Being literally rooted to the spot, most plants "bribe" animals into acting as couriers for their pollen, typically using food rewards. Plants that lack such rewards can deploy floral signals that mimic the food sources, mates, or oviposition sites of particular animals, thereby deceiving them into transferring pollen between flowers. Because of its function for pollination, floral mimicry is usually inviting (i.e. attractive to the operator), in contrast to animal mimicry, which is often (but not always) protective (i.e. deterring to the operator).

Mimicry in plants, and floral mimicry in particular, was once thought to be fairly improbable. Williamson (1982), for example, identified three aspects of plants that would hinder the evolution of mimicry. Firstly, their sessile condition would give operators greater time for discrimination and help them to learn the profitability of individual plants. Secondly, the aggregated distribution of most plants would reduce the likelihood of alternate exposure to models and mimics and allow operators to learn the profitability of patches. The third aspect, which is relevant only to protective mimicry, is the modular construction of plants, which means that the loss of any one relatively autonomous part, such as a branch, leaf, or fruit, is not as consequential for their fitness as it would be for animals. These three aspects are not absolute constraints on the evolution of mimicry in plants, particularly since the ephemeral nature of plant parts such as flowers and seeds would reduce the likelihood of location learning by operators, even more so when the plants themselves are highly dispersed, as is the case for many orchid mimics.

Despite these constraints, floral mimicry is widespread among plants. Food-source mimicry, including imitation of pollen, probably occurs in thousands of plant species. Attraction of flies and beetles to flowers which imitate their oviposition sites is a system that also involves thousands of plant species across a wide range of families and exploits the strong maternal instincts of the pollinators. Sexual mimicry is known to occur in hundreds of orchid species and has also been described recently in the daisy and iris families.

There are now literally hundreds of case studies of floral mimicry in plants, and we believe that this is an opportune time not only to showcase some of these studies but also to integrate their findings into a broad conceptual framework that can guide future research in the field. This publication is intended to be both a definitive book on floral mimicry and a broader treatise on the topics of floral adaptation and plant evolution.

Plants have typically been the Cinderella subjects of mimicry research, largely ignored in major treatments of mimicry. It needs to be appreciated that the disciplines of botany and zoology tended to develop separately prior to their merger into departments of biology or ecology and evolution, a process that started in the 1960s and which continues today. The separate development of these fields meant that zoologists did not always fully appreciate the importance of plant examples and that botanists did not always fully appreciate the need to study animal cognition and behavior in order to understand plant mimicry. An important question, posed in the introduction to this book and pursued as a theme throughout, is whether the evolutionary and ecological principles that were developed for defensive mimicry in animals also apply to floral mimicry in plants. Can we conceptually unify research on animal and plant mimicry and thereby overcome the past balkanization of the field of mimicry research along taxonomic lines?

To be fair to the zoologists who have written about mimicry without referring to plant examples, research on floral mimicry only started in earnest in the 1980s, although there were a number of key contributions prior to that time (Kullenberg 1961; Vogel 1978). The body of research on plant mimicry has been growing rapidly since then and now accounts for around 30% of all publications dealing with mimicry in multicellular organisms (Fig. 1.1). The vast majority of these plant-orientated papers deal with floral mimicry.

The discovery of floral mimicry

The discovery of floral mimicry preceded the work of Bates (1862) on butterfly mimicry by more than half a century and can be traced back to the classical writings of Christian Sprengel (1793) in his book *Das entdeckte Geheimnis der Natur im Bau und in der Befruchtung der Blumen* [*Discovery of the Secret of Nature in the Structure and Fertilization of Flowers*]. Sprengel's interpretation of the flower of the South African succulent *Stapelia hirsuta* that had been introduced into cultivation in Europe was that it "stinks like carrion only in order to lure bottle and carrion flies to which the stench is highly agreeable, and seduce them to fertilize the flower" (translation by Vogel 1996). Admittedly, Sprengel's commentary on floral mimicry was brief and lacked the detailed explanations that accompanied Bates's work, but did include the seminal realization that the function of certain signals in a mimic is to dupe an animal into misclassifying the mimic as something that it is not. The early discovery of carrion mimicry may also be owed to the fact that humans have acute perception of rotting flesh volatiles, so the association between the smell of a *Stapelia* flower and that of rotting flesh was easily made. Another very early (and vivid) description of carrion mimicry was in *The Temple of Flora* by Robert Thornton (1807). Thornton describes the flower of *Stapelia hirsuta* as having "something of an animal appearance" and "so strong a scent, resembling carrion, that blowflies in abundance hover round it; and mistaking the corolla for flesh, deposit there their eggs, which are soon converted into real maggots, adding to the horror of the scene, some being seen writhing among the purple hairs of the flower, and others already dead for want of food, the vegetable in this rare instance deceiving and overcoming the animal creation." The artist commissioned by Thornton tried to emphasize the sinister nature of the interaction, and perhaps the Medusa-like appearance of the *Stapelia* by including the head of a snake with its forked tongue protruding (Fig. 1.2).

In his original description of *Rafflesia arnoldii*, Robert Brown (1822) reproduced part of a letter from Dr. James Arnold, describing his first encounter with the world's largest flower: "When I first saw

(a)

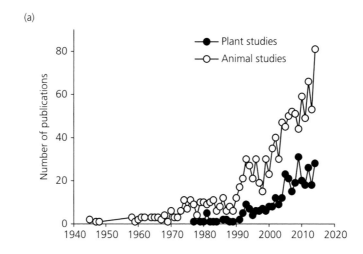

(b)

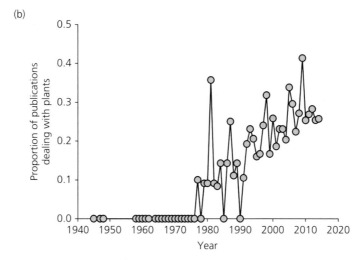

Figure 1.1 Relative growth in the number of publications dealing with mimicry in animals and plants, respectively, based on searches conducted in the Web of Science (© Thompson Reuters). (a) The number of publications indexed per year. These values are a minimum estimate due to limited indexing of the contents of earlier papers and books. Wickler (1968) estimated that 1500 articles on animal mimicry were published between 1860 and 1960. (b) The proportion of indexed publications on mimicry that deal with plant examples.

Figure 1.2 Illustration of the "maggot-bearing stapelia" (*Stapelia hirsuta*) from *The Temple of Flora* by Robert John Thornton (1807).

it a swarm of flies were hovering over the mouth of the nectary, and apparently laying their eggs in the substance of it. It had precisely the smell of tainted beef." Brown was aware of Sprengel's book and it was probably for this reason that he realized the significance of Arnold's observations. The first explicit use of the word mimicry in relation to flowers appears to be a letter from a schoolteacher, F. E. Kitchener, to Darwin in 1867 (Kitchener 1867). Kitchener asked Darwin whether anyone had "yet investigated the fertilization of the Stapelia, to see, whether the putrid smell may be regarded as a mimetic resemblance to carrion, which benefits the plant by attracting flies, under false pretences of its being a suitable place to lay their eggs." It seems possible Kitchener had already realized that Bates's by then well-publicized discoveries of mimicry among butterflies in the Amazon had important implications for our understanding of floral traits.

Sexual mimicry was discovered in 1917 by Maurice-Alexandre Pouyanne, who had observed male scoliid wasps grasping at flowers of *Ophrys speculum* in Algeria and interpreted this as an attempt at copulation triggered by visual similarity between the flowers and female wasps (Pouyanne 1917). Not long afterward similar systems were discovered in Australia by Edith Coleman, who published a series of papers on the pollination of *Cryptostylis* orchids by sexually aroused ichneumonid wasps (Coleman 1930). Despite its early discovery, sexual mimicry was largely ignored (or simply not accepted) by mainstream natural historians until detailed observations by John Godfery and Bertil Kullenberg, in particular, broadly established the phenomenon in botanical and zoological textbooks (Godfery 1925; Kullenberg 1961).

The deception of food-seeking animals by flowers that do not offer rewards was first identified by Sprengel (1793), but these cases involved generalized deception rather than mimicry of particular species. The discovery that some plants mimic specific food plants of their pollinators was made only during the 1960s, with a number of possible examples being suggested by van der Pijl and Dodson (1966). The reason for this, we suspect, is that unlike brood-site mimicry and sexual mimicry which are associated with highly derived phenotypes, food-source mimicry is easily confused with floral

syndromes and requires an analysis of the interactions between plants and pollinators in a community context.

Key concepts

Adaptive resemblance

The literature is replete with putative examples of mimicry that have never been properly tested. What are the basic conditions that must be met for a resemblance between species to be considered an example of mimicry? Firstly, and most importantly, there must be evidence that the resemblance of one species to other species or their by-products must have arisen through selection for resemblance per se (Starrett 1993). This excludes incidental resemblance, as well as resemblance due to convergent evolution in which unrelated species adapt to common functional requirements.

Adaptive resemblance is a fundamental criterion for all cases of mimicry and yet is very seldom tested explicitly. It is possible to use experiments to test whether individuals that more closely resemble a model species have higher fitness than those that less closely resemble the model (Jersáková et al. 2012; Newman et al. 2012; Pfennig et al. 2015), or to test whether a mimic performs better when it occurs with its model than when it occurs without its model (Peter and Johnson 2008; Ries and Mullen 2008). Such experiments demonstrate that traits that confer resemblance have a function. However, hypotheses about adaptation also require testing in a historical framework by asking whether the traits that confer resemblance are indeed evolutionarily novel within a lineage (Harlin and Harlin 2003; Johnson et al. 2003; Ma et al. 2016). This can be achieved by mapping the occurrence of these traits onto the phylogeny of a group of closely related organisms that includes mimics and non-mimics.

Cognitive misclassification

A basic assumption about mimicry is that the operator should consider the mimic to be a representative of the model. Cognitive misclassification is sometimes obvious from the behavior of the operator. A male insect that attempts to copulate with a flower, for example, has clearly misclassified that flower as being a female insect. Cognitive misclassification is less easy to determine in cases where an operator does not show a stereotyped behavior associated with a model. When a predator avoids a palatable insect that resembles an unpalatable insect there is a possibility that the predator was showing a general innate avoidance of particular signals and not confusing one species for the other. Similarly, when a pollinator probes non-rewarding flowers that resemble the rewarding flowers of a different species, how do we determine whether there has been actual misclassification? One option is to test whether the probability of repulsion of an attack (in the case of protective mimicry) or attraction (in the case of floral mimicry) increases with the presence of the model. If so, the argument for cognitive misclassification is considerably strengthened. To qualify as a mimic, an organism must exploit the behavioral responses of operators to the specific phenotypes of model organisms.

A criticism that could be leveled at the concept of cognitive misclassification is that it implies a process of categorization by signal receivers. Another way of approaching the problem is to think of it in terms of receiver errors. According to signal detection theory, the receiver does not need to consider the mimic to be a representative of the model, but merely needs to accommodate the possibility of error in its sensory systems and set an acceptance threshold for appearance based on the most adaptive ratio of models: mimics. When the cost–benefit ratio of a system changes, so too would the acceptance threshold (Oaten et al. 1975).

Advergent evolution

Contrary to a common misconception, the resemblance between most mimics and their models is not usually due to convergent evolution, because the resemblance between the two species does not arise from independent adaptation to a common environmental factor. Instead, the resemblance arises from the effect of the model on the traits of the mimic. Consider the classic example of Batesian protective mimicry, where a palatable species mimics the signals of an unpalatable species. This pattern of evolution, where a mimic evolves

a resemblance to its model while the model itself either does not evolve at all or, if it is harmed by the interaction, is under selection to diverge from the mimic because of the cost of parasitism, is known as advergent evolution (Brower and Brower 1972). Advergence applies to most mimicry systems, including floral mimicry (Fig 1.3). A particularly obvious example involves floral mimicry of non-living models such as carrion and feces.

Müllerian mimicry (adaptive resemblance among unpalatable prey or rewarding plants) was thought to represent a pattern of convergence arising through coevolution, but historical reconstruction of Müllerian mimicry systems suggest that younger species may mimic the signals of older species, thus resulting in a pattern of advergent evolution (Hines et al. 2011). Furthermore, variation in the abundance of species and palatability (in the case of flowers, rewards) among species makes it likely that Müllerian systems contain elements of advergence (Mallet 1999; Sherratt 2008). The reality is that many mimicry systems are probably "quasi-Batesian" (Speed

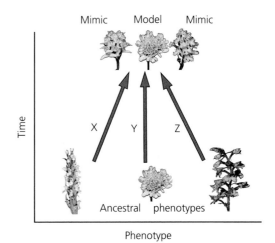

Figure 1.3 Patterns of adaptive phenotypic evolution in a floral food-source mimicry system. In this example mimicry of the phenotype in lineage Y evolves independently in lineages X and Z. The modifications for mimicry in X and Z result in a pattern of advergent evolution relative to the phenotype in Y, which does not change over time. In addition, the resemblance that develops between X and Z is a pattern of convergent evolution in response to a common selective factor (mimicry of the phenotype in the Y lineage). Adapted from Brower and Brower (1972) and Johnson et al. (2003). Photos © Steve Johnson.

1993) and combine the properties of Batesian and Müllerian mimicry.

Honest signals and reliable cues

The models in most cases of animal mimicry are other animals that have conspicuous signals denoting that they are dangerous or unpalatable. Edward Poulton (1890) coined the term aposematism to describe this form of honest signaling. Aposematism is usually associated with bright colors, but the strategy can include odors, sounds, and even behavior. Several types of signals can be combined in a multimodal display. Predators may have an innate aversion to aposematic signals, but more commonly the strategy works via a process of predator conditioning, often referred to as its education. Animals learn to avoid the signals of aposematic species, so the more individuals that are present in a population the more likely it is that any one individual will gain protection as a result of aposematism.

Floral advertising is functionally similar to aposematism in that it involves the deployment of signals that maximize conspicuousness to signal receivers. More than 90% of all angiosperms produce flowers that contain food rewards, and these flowers advertise that they are rewarding. Like aposematism, floral advertising usually involves honest signals, but, unlike aposematism, these signals serve to attract the signal receivers. Pollinators may have a degree of innate attraction to these signals, but more commonly they learn to associate the signal with rewards. Therefore, in the same ways that palatable animals can exploit animals conditioned by the signals of aposematic animals, non-rewarding flowers can exploit animals conditioned by the signals of rewarding flowers.

One idea behind honest signaling is that it should act as a handicap, and thus be unaffordable for low-quality individuals, and hard to imitate. This principle probably applies to flowers: because larger plants can afford larger floral displays, larger flowers generally have more nectar, and pollinators usually respond positively to larger floral displays. Alternatively, honest signals can evolve through the punishment of cheaters if plants that signal dishonestly are visited less or totally avoided by pollinators. Punishment of cheaters is most relevant for

plants that require frequent repeat visits by pollinators to achieve maximum fecundity.

Sexual mimicry in flowers is based on the exploitation of a different kind of honest signal, the sexual displays of female insects. Sexual displays are usually honest (i.e., representative of the value of the mating partner) because signals that are not honest would lead to mating with inferior partners and thus be less likely to spread in a population. The sexual signals of insects are not only exploited by plants, but also some spiders that imitate the pheromones of moths in order to attract prey (Eberhard 1977; Stowe et al. 1987).

An important exception to the general rule that mimicry involves exploitation of honest signals is the case of oviposition-site mimicry in plants. Here the model is often a dead animal or feces, and the volatiles they emit are not signals per se but a byproduct of the process of decomposition. While these volatiles are not signals, they do serve as reliable cues for insects to locate their oviposition sites. One interesting exception is rotting fruit which serves as a model for certain floral mimics. The volatiles of rotting fruit are mostly derived from yeasts. They have historically been considered a by-product, but there is a recent view that these volatiles may, in part, be an honest signal by yeasts as part of their strategy for dispersal by fruit flies and other insects (Becher et al. 2012).

Imperfect mimicry

Mimics often possess a rather generalized resemblance to their models. In many cases it is not even possible to identify a specific model, but only a broad category of models. Hover flies, for example, often resemble wasps or bees, but only in a rather generalized sense. This phenomenon is known as imperfect mimicry, and has seldom been considered in the context of floral mimicry. Many of the systems that we discuss in this book can be considered as examples of imperfect mimicry. Orchids that deploy generalized food deception are often pollinated by a specific pollinator functional group such as bumblebees, megachilid bees, or cetoniine beetles. These orchids probably attract pollinators because of their generalized resemblance to rewarding plants pollinated by these insects (see Chapter 3). Flowers that mimic insect oviposition sites often deploy a generalized set of signals that insects would associate with their oviposition sites, but are not mimics of a specific model species (see Chapter 6). *Dracula* orchids, for example, seem to generally mimic the mushroom oviposition sites of flies and not a specific mushroom species (Policha et al. 2016).

Several possible explanations for imperfect mimicry have been identified (Kikuchi and Pfennig 2013):

1. *Eye-of-the-beholder*. The first and most obvious explanation is that while a mimic may seem imperfect to a human observer, it may nevertheless be effective in terms of fooling an operator into classifying it as an example of its models.
2. *Breakdown of mimicry*. Traits of a mimic that extends beyond the range of its model or occupies a region in which the model is rare may no longer be maintained by selection.
3. *Multiple models*. A generalized form of mimicry may allow a mimic to gain fitness in different contexts with different models (Edmunds 2000). A related explanation is selection from multiple operators, as in the case of non-rewarding plants that attract a broad assemblage of pollinators with different signal preferences and different degrees of conditioning. We think that multiple models and multiple operators are likely to apply to many non-rewarding plants that deploy generalized food deception.
4. *Relaxed selection*. Studies of hover flies have shown that larger species are more likely to be accurate mimics than smaller species (Penney et al. 2012). This is probably because larger individuals are more attractive to predators. The analogue in plants is that species with food rewards are under less selection to develop mimicry than are species that lack rewards.
5. *Developmental constraints*. By virtue of their Bauplan, certain plants and animals are unlikely to ever be perfect visual mimics of certain models.
6. *Receiver bias*. Floral sexual mimics can be more attractive to male insects than their female insect models, yet deviate significantly in attractive

signals. This hypothesis implies that it is the models, not the mimics, that are most constrained in terms of developing the optimum phenotype.

Imperfect sexual mimicry in orchids may be related to the sensory preference of male insects living in inbred colonies for females that have novel chemical phenotypes (Vereecken and Schiestl 2008). Males may exhibit this preference because novel chemistry is an indicator of genetic distance, and mating with females that have a novel scent may result in offspring that are fitter because they are outbred (Vereecken et al. 2007). Imperfect floral sexual mimics are involved in highly specialized interactions (a single model and a single pollinator) and demonstrate that imperfect floral mimicry is not necessarily related to multiple models or breakdown of mimicry.

Another factor that should be borne in mind when considering imperfect mimicry is that predators and pollinators have to make speed–accuracy trade-offs (Chittka and Osorio 2007). Even if they are ultimately capable of distinguishing a mimic from the model, they may seldom do so in nature because the time required would be prohibitively expensive in terms of energetics. This means that a generalized resemblance to a model or set of models may be optimal for fitness, particularly if more accurate mimicry involves trade-offs against factors such as thermal balance or growth.

Accurate mimicry

Imperfect mimicry is actually a misnomer, because it implies that other mimics are perfect. Since mimicry can never be perfect, we prefer to contrast imperfect mimicry with accurate mimicry, and acknowledge that these are simply points along a continuous spectrum of possibilities.

There are two ways to measure the degree of refinement (accuracy) in a mimicry system. The first is to compare the signals of mimics and models using tools that simulate the perceptual system of the operator (see Box 3.1). The second is to conduct experiments with the operator and work out the extent to which operators are duped by the mimicry. In the previous section (Imperfect mimicry) we raised the theoretical possibility that a mimic that appears imperfect to humans might in fact be highly successful in duping an operator. For this reason we favor behavioral approaches to the study of mimicry, specifically experiments in which operators are presented with choices between putative mimics and their models. Such experiments can be used to determine whether operators show discrimination and also whether their behavior toward the putative mimic is similar to that observed toward the model.

Using this approach, it has been shown that pollinators often do not discriminate between non-rewarding floral food source mimics and their rewarding models. In many of these systems the probability of a pollinator visiting a mimic versus a model when it is offered an equal choice has been shown not to deviate statistically from 50:50 (Johnson 2000). However, it is not a strict requirement for mimicry that an operator be completely unable to discriminate between mimics and model, only that it should have difficulty in discrimination. There is another reason to be cautious about discrimination experiments. Lack of discrimination between models and mimics could simply indicate that the operator is exploring new food sources.

Frequency dependence

Mathematical simulations of Batesian mimicry invariably demonstrate that mimics benefit when there is a high ratio of models to mimics (Ruxton et al. 2004). This is because these simulations assume that the responses of the operator are conditioned by the frequency of encounters with models and mimics. Indeed, associative learning is central to theoretical models and empirical studies of protective mimicry (Skelhorn et al. 2016). However, there is no absolute requirement that models should outnumber mimics. If operators have poor sensory discrimination and/or learning ability, mimics may be able to perform well even if they greatly outnumber their models. The degree to which the mimic resembles the phenotype of the model is, of course, equally important for determining how quickly operators would learn to discriminate between the organisms. The prediction that mimics should perform best when they are rare relative to their models has been supported by a limited number

of studies of floral food-source mimicry systems (Anderson and Johnson 2006). Sexual mimicry is an exception to this general prediction of Batesian mimicry because the pollination success of mimics may actually decrease when the female hymenopteran models emerge.

The absolute number of individuals of mimics and models is also important for the learning process in mimicry systems. Batesian mimics would be expected to perform less well as their absolute number increases. This is well documented in sexual mimicry systems and has been attributed to a learning process in male hymenoptera (Peakall 1990; Ayasse et al. 2000). The extent to which this would disadvantage the mimics would again relate directly to the learning and sensory discrimination ability of the operators. Studies that test the effects of both abundance and mimic–model ratios on mimic performance are extremely rare (Roy 1996; Ruxton et al. 2004).

There has been some confusion over the meaning of "frequency dependence" in mimicry systems. It is particularly important to distinguish between the ecological effects of the relative and absolute frequencies of the mimic and model organisms, as discussed above, and the population genetic concept of "frequency-dependent selection" whereby the fitness of a morph varies according to its frequency in a population. There are several examples of animal Batesian mimics that exhibit polymorphisms (often sex-linked) within populations (Turner 1977). A common explanation for the maintenance of these polymorphisms is that the morphs mimic different models in the same community and are maintained by predators discovering the palatability of common morphs and/or avoiding morphs that are rare relative to the models. Roy and Widmer (1999) argued that a similar process of negative frequency-dependent selection in floral Batesian mimicry systems would result in oscillation of morph frequencies over time, and therefore lead to high levels of phenotypic variation in mimics. However, Batesian mimicry can also result in fixation of those morphs that are the most effective mimics of a locally abundant model, and thus lead to low levels of phenotypic variation in signaling traits within mimic populations (Joron and Mallet 1998). Another consequence of this process of fixation is that floral mimics can exhibit geographic polymorphisms arising from mimicry of different models across their geographic range (Johnson 1994).

The effectiveness of Müllerian mimicry systems in the ecological sense should increase with the absolute numbers of all participating species. In the population genetic sense, Müllerian mimicry should be characterized by positive frequency-dependent selection, but different morphs could become fixed in different populations leading to geographic polymorphisms.

Negative frequency-dependent selection is expected to be important in generalized food deception, as pollinators would be expected to learn to avoid more common morphs, but the evidence for this so far is rather equivocal (see Chapter 3).

Another issue of relevance to the frequency-dependent success of floral mimics is that they usually pay a lower cost when mimicry is ineffective, i.e., when operators are not duped by the mimicry. In protective mimicry, a palatable animal is likely to lose its life if the mimicry system fails, while a plant may merely lose out on a pollinator visit. Given that plants are long-lived, they may still benefit from mimicry even if the system is not very effective and pollinators are duped only occasionally. For this reason, predictions about frequency dependence based on modeling of demography and trait evolution in protective mimicry may not apply strictly to plants.

Predictions for mimicry systems

Apart from the fact that protective mimicry in animals results in the repulsion of operators and floral mimicry in plants results in their attraction, the predictions that can be made for mimicry systems in animals and plants are very similar. These are that:

1. Mimics and models should occur in the same habitat, have overlapping phenologies, and interact with the same operators.
2. The mimic should resemble the model to the extent that operators have difficulty in discriminating between them.
3. The responses of operators to signals of the mimic should be shaped by their experience with the model.

4. The fitness of the mimic should be higher in the presence of the model than in its absence.
5. Individuals that resemble models more closely will experience higher fitness than individuals that resemble models less closely.
6. Mimics should perform best when models are relatively abundant.

And, as a consequence of all of the above:

7. The geographic distribution of the mimic should be nested within that of the model.

In addition to these core predictions, we can make the additional prediction:

8. In the case that mimics do extend beyond the range of a single model, they should exhibit geographic variation in signaling traits, either because they mimic different models in different geographic regions, or because of the loss of mimicry outside the range of the original model.

The semantics of floral mimicry

Batesian and Müllerian mimicry are named after the nineteenth-century naturalists Henry Walter Bates and Fritz Müller, who described mimicry of the wing patterns of unpalatable butterflies by palatable butterflies, and wing pattern mimicry among unpalatable butterflies, respectively. The two concepts have been applied to protective mimicry in a wide range of organisms and have largely stood the test of time, although it is now widely accepted that there is a continuum between Batesian and Müllerian mimicry, corresponding to a palatability gradient (Speed and Turner 1999). In addition, Müllerian mimicry systems are usually assembled by the consecutive addition of new species, such that some species are original models and others mimics (Hines et al. 2011), and not by coevolution from the same historical starting point, as was previously believed. These debates about the distinction between and essential nature of Batesian and Müllerian mimicry continue to take place within the field of protective mimicry (Ruxton et al. 2004). Of special interest to us here is whether these concepts should be broadened to include floral examples.

There has been considerable disagreement over the application of the concepts of Batesian and Müllerian mimicry to pollination systems. Some authors, such as Ford (1975), Wiens (1978), and Little (1983), argued that the terms were not appropriate for floral mimicry systems on the grounds that operators are repelled in protective mimicry systems and attracted in floral mimicry systems. Others, including Brown and Kodric-Brown (1979) and Roy and Widmer (1999) have used the terms Batesian mimicry to describe any non-rewarding flowers that rely on mimicry to achieve pollination and Müllerian mimicry to refer to adaptive resemblance among rewarding flowers.

Dafni (1984) argued that the concept of Batesian mimicry should only be applied in cases where non-rewarding plants use mimicry to exploit the behavior of pollinators conditioned on food sources, but not in cases such as sexual and oviposition-site mimicry where the original attraction to the mimic is based solely on innate responses. However, it is difficult to draw a clear distinction between innate and learned responses, even in the case of protective mimicry systems. There is now evidence for innate responses of pollinators to signals of food flowers, and this could be exploited by floral food-source mimics (Dötterl et al. 2011; Milet-Pinheiro et al. 2012). Furthermore, while the initial attraction of male insects to sexual mimics may be innate, operators can learn to avoid sexual mimics, either on an individual basis or by leaving an area where sexual mimics flower, after a visit (Schiestl 2005). Another solution to the conundrum was offered by Endler (1981) who grouped several related mimicry phenomena, including floral mimicry (which he refers to as reproductive mimicry), under the category of "Batesism." This recognizes the conceptual overlap between Batesian protective mimicry and most cases of floral mimicry, yet allows for Batesian mimicry to be used in its historical sense as a form of protective mimicry.

Given that we can never be certain about the relative importance of innate and learned preferences for the evolution of floral traits and because we wanted to avoid a complex classification of special botanical terms, such as "Dodsonian" and "Bakerian" mimicry (Pasteur 1982), we have followed Roy and Widmer (1999) and applied the terminology of Batesian mimicry to all examples of non-rewarding flowers that mimic specific models and thereby exploit the behavioral responses of operators toward the model phenotype. This is a purely pattern-based classification; we acknowledge that the properties of mimicry, including frequency dependence, and even the processes of evolution, are likely to differ among, and even within, food-source, oviposition-site, and sexual mimicry systems.

The use of a pattern-based classification becomes more problematic when we consider Müllerian mimicry. We cannot consider all examples of floral resemblance among rewarding species that share pollinators to be Müllerian mimicry because resemblance among rewarding flowers (and even among non-rewarding and rewarding flowers) can arise from the evolution of floral syndromes. Floral syndromes are patterns of convergent evolution among flowers that adapt to the sensory preferences and morphology of common pollinators. This is a different scenario from Müllerian mimicry, in which the traits of species co-evolve because the resemblance to each other enhances their fitness. Consider the simple case of a community of red flowers pollinated by birds. If these flowers are red because this color is more conspicuous to birds than any other color, then the red color reflects a syndrome. If the community of plants has red flowers because species benefit from sharing the same flower color as other species, then Müllerian mimicry may apply. It can be very difficult to distinguish between these two scenarios. One method is to test whether putative Müllerian mimics have reduced reproductive success when co-mimics (Müllerian mimics are both models and mimics) are removed from a community, while in the case of species that merely share a syndrome, removal of one species should have little effect on reproductive success, or even increase it if there is competition among species. Müllerian mimicry should also lead to geographic polymorphisms, while syndromes should be relatively consistent among communities. The application of Müllerian mimicry to floral systems is discussed further in Chapter 7 which is devoted to special and unresolved cases.

Exploitation of receiver bias

Receiver bias is an important aspect of the evolution of generalized food deception and floral mimicry. The general term "receiver bias" used here includes phenomena such as perceptional bias, pre-existing bias, and sensory traps, which all have slightly different meanings but generally refer to the same principle (Endler and Basolo 1998). In its broad sense, receiver bias can be defined as selection imposed by an interacting organism through sensory or perceptional properties that have evolved in a context other than the actual interaction. For example, male insects visiting flowers may select for sexual signals because they have evolved an extraordinary sensitivity as well as innate preferences for them. Receiver bias can also apply to learning, for example when social bees have a special predisposition to learn certain signals that also play a role in their social interaction within the hive. Viewed as a plant strategy, exploitation of perceptional biases (EPB) (Schaefer and Ruxton 2009) can be helpful for understanding mimicry-like phenomena as well as imperfect mimicry since EPB may select for optimal mimicry phenotypes that differ considerably from that of the model. It is also a useful concept for understanding the evolution of certain signals that are part of generalized food-deception strategies which do not involve specific models. It is likely to apply to rewarding systems, too, such as the evolution of large flowers (unless size is an honest signal), dark or yellow spots on flowers in plants pollinated by bees that otherwise collect pollen, or the emission of green leaf volatiles to attract hunting social wasps as pollinators. Because receiver bias is an important process in rewarding, generalized deceptive, and mimetic pollination systems, it is not surprising that many intermediate forms between these pollination systems exist, making it hard find a clear-cut definition for each phenomenon. The problem is that when the "model" is defined too widely, virtually every system where receiver bias shapes the evolution of floral signals will fit the definition of mimicry. For example, are "green-leaf volatiles" emitted after an attack by herbivores an appropriate model for a definition of mimicry? Such volatiles are emitted by the flowers of figworts (*Scrophularia*) to attract hunting social wasps as pollinators (Brodmann et al. 2012). Or is a nest entrance a good model for a potential mimicry system, involving flowers with dark centers and stingless bees that have innate preferences for these patterns (Biesmeijer et al. 2005)? Similar questions can be asked for large flower size, pollen, general floral volatiles, etc. Trying to sort out "real" from "non-real" models can easily become arbitrary and yet is somehow unavoidable. In our view, receiver bias is a key process in the gradual evolution of mimicry, and it helps to explain the frequent occurrence of intermediate systems, with "fuzzy" models, that are sometimes rewarding, and why mimics are sometimes able to survive outside of the range of their models.

There has been disagreement over the role of receiver bias in the evolution of accurate mimicry systems (Ruxton and Schaefer 2011). Sexual mimicry in flowers is undoubtedly accurate (it often stimulates copulation behavior by male insects), yet it probably evolves almost entirely through the exploitation of receiver bias. We feel that the disagreement over the role of receiver bias is related to whether mimicry is used to connote a process or a pattern. Schaeffer and Ruxton's (2009) dichotomy between receiver bias and mimicry makes sense if mimicry is viewed as an evolutionary process involving selection by operators that are conditioned on signals of models, while receiver bias does not require experience with models or precise matching to the actual signals of models. However, there is no doubt that exploitation of receiver bias has been important in the evolution of the signals of flowers in general and the signals of floral mimics in particular.

Mimicry versus deception

Acknowledging the importance of receiver bias in the evolution of mimicry makes it obvious that floral mimics do not always have to be deceptive in the sense of not offering a reward. Obviously, mimicry usually involves some form of "pretence" (i.e., a false promise), which is already implied in the meaning of the Greek word *mimos*, "imitator, actor." Therefore, pollinators often do not find in mimetic flowers what they originally searched for, but may still find an appealing reward. For example, two color morphs of *Turnera* (Turneraceae) which resemble co-flowering mallows (Malvaceae) are visited by bees specialized on the pollen of Malvaceae that are obviously searching for mallows, yet actively collect *Turnera* pollen (Benitez-Vieyra et al. 2007). Whereas this seems to be a clear-cut example of a rewarding (Müllerian) mimicry system (unless *Turnera* pollen is malnutritive or non-digestible for mallow bees), many intermediate systems exist. For example, oviposition mimics sometimes produce nectar, while attracting the pollinator with false signals of a suitable substrate for egg-laying (see Chapter 6). Although the nectar may be readily consumed, there is often a strong cheating component imposed by these mimics, including imprisonment and infanticide. The question of whether a mimicry system is rewarding or deceptive can be tricky, and depends on the definition of deception. *Epipactis helleborine*, for example, has been classified as deceptive despite its abundant production of nectar, because the flowers attract predatory social wasps with green leaf volatiles that the wasps seemingly associate with the presence of the feeding caterpillars that constitute their prey (Brodmann et al. 2008). One solution to the deception/reward question may be offered by the behavior of the pollinator, or the fitness consequences of visiting a flower. When pollinators show negative associative learning, and avoid mimics after visits, deception can be assumed. This is obvious in sexual mimics, where pollinators quickly avoid individual flowers (Ayasse et al. 2000) or locations with flowers (Peakall 1990) after deceitful visits (see Chapter 5). If a pollinator continues to make frequent visits to flowers of the mimic, its flowers may actually be rewarding; however, some pollinators may simply be unable to avoid a non-rewarding mimic due to constraints in their learning ability. Potential negative fitness consequences of visitation to a mimic may also help to decide on the deceptive nature of a pollination system, but will typically be difficult to assess in nature.

Modalities of signaling

Visual, olfactory, and tactile signals can all be important in floral mimicry systems, but often to differing degrees. The traditional focus has been on visual cues, probably because they are prominent in our own human perception of the world. We now use advanced models of animal vision to analyze the extent to which signals of mimics and

their models can be distinguished by their pollinators (see Box 3.1). However, it is well known that some forms of mimicry, notably sexual and oviposition-site mimicry, are largely based on chemical cues. Indeed, one of the major aims of this book is to review recent advances in the chemical ecology of mimicry systems. One reason why floral mimicry has taken longer than animal mimicry to develop into a fully fledged field of research is that the tools for deciphering the chemical language of plants and for studying the chemical ecology of biological interactions have only been developed relatively recently. An interesting aspect of mimicry is that it usually involves multimodal signaling. For example, although chemical signals are usually sufficient to attract pollinators of sexual mimics, and often also to induce them into landing and performing some-pre-copulatory behavior, sexual mimics usually also deploy some of the olfactory, visual, and tactile signals or cues of their models (Gaskett 2011). Similarly, in carrion mimicry a combination of volatiles associated with the oviposition substrate and a dark target has been shown to be most effective for inducing landing behavior by certain flies (Brodie et al. 2014).

Niches and ecological specialization

Ecological specialization is often a result of adaptation for a particular available niche. Specialization in the pollination systems in plants is usually a result of specialization for particular pollinator niches (Johnson 2010). For example, plants may evolve white scented flowers with long floral tubes to exploit hawk moths as pollinators (Johnson and Raguso 2016). Floral mimicry systems represent a further level of specialization, not only to a particular group of animals, but also to a particular aspect of the biology or life history of the animal. As a general rule, the evolution of accurate mimicry in flowers is associated with ecological specialization in the pollination system. For example, flowers that deploy generalized food deception usually have much broader assemblages of pollinators than do plants that are accurate mimics of specific food plants.

There is still uncertainty about the levels of specialization in oviposition-site mimicry. In the case of carrion mimicry, for example, studies suggest that carrion insects often specialize on different stages of degradation of the carcass, making it likely that carrion flowers use signals of a particular degradation stage to attract particular insect pollinators.

One reason why mimicry acts as a driver of diversification of plants is that there are a very large number of mimicry niches that can be exploited. For example, the mating system of almost every insect species in a community is a niche that can be potentially exploited by a sexually deceptive plant species. Similarly, insects have a vast range of oviposition sites representing potential niches for exploitation by plants. Many of these niches are in fact used by flowers, but the number of niches that are not used is far greater than the number that are.

The basis for evolutionary specialization is the trade-offs that exist between adaptations to one niche and those for another niche. These trade-offs are particularly evident in mimicry systems. It would not be easy for a plant to mimic a bird- and a moth-pollinated flower at the same time, as the signals that attract birds and moths are very different (Aigner 2001). Similarly, sexual mimicry in flowers usually involves the exploitation of the mating system of a single hymenopteran species, because the signals that are used in insect mating are themselves highly specific. A blend of signals may be expected to work less well than a "pure" signal.

There is another level of specialization evident in floral mimicry systems: Floral mimics tend to exploit interactions which are already specialized, such as the mating system of a single insect species or an interaction between a pollinator species and its food flowers, or between a particular group of insects and their oviposition sites. These specialized interactions in nature can be exploited by mimics because they involve the use of very specific cues (or signals) by the operators. A general rule about floral mimicry is that it involves cues that an operator cannot afford to ignore. These might be the cues of a specific food plant, a female mating partner, or an oviposition site. Floral mimicry works as an evolutionary strategy because operators would suffer massive losses of fitness if they were to ignore these cues, or be too choosy about to whether or not to respond to them.

In general, available data suggest that the targeting of very specific operators is a fairly unique feature of floral mimicry. Most cases of protective mimicry in animals target a much broader set of operators involved in predation.

The structure of this book

The chapters of this book are designed to highlight particular systems of floral mimicry and to integrate them into the broader theory of mimicry. We have looked for common ground between mimicry in plants and animals, but also acknowledge some of the biological differences between these two groups in terms of how mimicry evolves and functions.

We begin in Chapter 2 by examining the evolutionary factors behind the losses and gains of floral rewards. We emphasize that while mimicry is associated with plant lineages that lack floral rewards, rewardlessness and mimicry are not synonymous. Rewardlessness usually evolves before mimicry and the selective factors for these two strategies often differ.

In Chapter 3 we focus on the phenomenon of generalized food deception, a highly successful strategy by which rewardless species attract pollinators without relying on mimicry. Generalized food deception is usually the precursor for the evolution of floral mimicry, particularly in the orchids.

The focus of Chapter 4 is food-source mimicry in non-rewarding flowers and its close parallels with Batesian protective mimicry in terms of frequency dependence and the evolution of geographic polymorphisms. Some aspects of food-source mimicry are controversial and Chapter 4 is organized around the different lines of evidence for this form of mimicry in plants.

In Chapter 5 we deal with sexual mimicry, which involves extensive floral modifications, particularly the evolution of highly specialized volatile signals to attract male insects. The study of sexual mimicry in plants has given new insights into insect reproductive biology and provides some of the best examples of pollinator-mediated plant speciation.

In Chapter 6 we focus on the widespread phenomenon of floral oviposition-site mimicry. This includes mimicry of fruit, carrion, feces, and fungal substrates that are used by insects as brood sites. Chemical signals are key in all of these systems, but there is also evidence for combined effects of chemical and visual signals (i.e., multimodal signals).

In Chapter 7 we consider special cases of mimicry, some of which are contentious. The concept of Müllerian mimicry has been applied to resemblance among rewarding flowers, but we show that the evidence that this reflects a process of mimicry remains tenuous. Resemblance between the male and female flowers of dioecious or monoecious plants only represents mimicry when it arises from selection for similarity per se, particularly selection for similarity among non-homologous features of the sexes. Such selection is particularly strong in cases where only one sex offers a reward.

Finally, in Chapter 8, we look to the future of mimicry research and focus on molecular approaches and their potential to answer unresolved questions about speciation. The inclusion of this chapter does not mean that we see the future of mimicry research as being purely molecular and not ecological. However, we felt that molecular approaches are not specific to a particular type of mimicry system, and it thus made more sense to deal with this topic in a single chapter and to discuss ecological approaches in the chapters that focus on each type of floral mimicry.

CHAPTER 2

The evolution and maintenance of floral rewardlessness

Introduction

Pollination typically takes the form of a biological transaction, whereby plants offer a material reward in exchange for a sexual service facilitated by animals. This is considered a mutualism because both partners benefit. In the public imagination, plant–pollinator relationships are often believed to involve cooperation, but in reality these systems involve intense mutual exploitation. Furthermore, either partner will readily "cheat" if provided with the opportunity to do so (Bronstein 2001). It is not uncommon, for example, for flower-visiting animals to rob flowers of their nectar by biting or poking a hole through the side of the corolla, thus avoiding the usual way in through the flower entrance which is needed for pollination (Irwin et al. 2010). Likewise, many plant species are non-rewarding and dupe their pollinators by deploying floral signals that suggest the presence of a reward that is not actually there (Renner 2006). In such cases, the relationship between plant and animal is one of parasitism. This chapter is devoted to an exploration of the prevalence, evolution, and functional significance of floral rewardlessness.

The rewards that plants offer to entice animal pollinators include food such as nectar and pollen, nest-building materials such as resins or oils (also used as food), and harvestable sex attractants such as fragrances. Other rewards include shelter, mating rendezvous sites, and brood sites in the form of ovules or—in the case of cycads—male cones that are eaten by developing larvae (Simpson and Neff 1983). Systems of floral deception are known for most of these reward systems, though some are

relatively rare. For example, we know of no cases in which pollinators seeking resins or fragrances are duped by plants—this may relate to the fact that these rewards also function as (honest) signals in these systems, making the deception of pollinators unlikely. By far the most common form of floral deception involves the duping of animals seeking nectar (Renner 2006).

As explained in Chapter 1, not all deceptive plant species rely on mimicry (some rely solely on generalized attractive signals), but there is certainly a strong association between floral rewardlessness and various forms of mimicry. A key question is whether rewardlessness generally evolved before, together with, or after the evolution of mimicry in plant lineages, and we return to this issue later in this chapter. Deception and rewardlessness are obviously not always synonymous, since wind- and water-pollinated plants (as well as many obligate selfers) have no rewards and do not deceive animals. It is also important to clarify that we use the term deception in relation to an absence of the rewards that are sought by the particular flower visitor. For example, flowers that lack nectar may deceive nectar-seeking animals, yet provide a pollen reward for pollen-seeking animals (Pellmyr 1986). *Stapelia* flowers that secrete small amounts of nectar deceive female flies seeking brood sites, even though flies may consume some nectar (Meve and Liede 1994). Similarly, male flies visiting the flowerheads of *Gorteria diffusa* are sexually deceived, even though these flowers produce small amounts of nectar (Ellis and Johnson 2010). These examples still qualify as deception according to the useful definition offered by Bradbury and

Floral Mimicry. Steven D. Johnson & Florian P. Schiestl.
© Steven D. Johnson & Florian P. Schiestl 2016. Published 2016 by Oxford University Press.
DOI 10.1093/acprof:oso/9780198732693.001.0001

Vehrencamp (2011) of deception as the "the provision of inaccurate information by the sender, such that the sender benefits from the interaction, but the receiver pays the costs of a wrong decision." It can be challenging to determine whether or not a flower is deceptive, because it requires natural history knowledge and sometimes experimentation to establish the motivation behind the responses of animals to floral signals, and thus whether or not they are duped in terms of paying a cost for the wrong decision. One way of identifying deception in pollination systems is by observing whether pollinators re-visit flowers or whether they avoid them in subsequent visits (see also Chapter 1). It is also important to note that we reserve the term deception for plants that employ a particular strategy and are thus genetically fixed for an absence of the rewards that are being sought, not those that are transiently rewardless because their floral rewards have been depleted by pollinators.

Darwin's "gigantic imposture"

Christian Sprengel (1793), considered by many to be the founder of pollination biology (Lloyd and Barrett 1996), was the first to realize that the flowers of most European *Orchis* species lack nectar; he referred to these as *Scheinsaftblumen* (sham nectar flowers). Darwin (1862, 1877) consistently refused to accept the idea that flowers could be rewardless and yet attract pollinators. He felt that this would represent a "gigantic imposture" that cannot be reconciled with evidence, for example the multiple loads of pollinaria on moths visiting the flowers of deceptive orchids such as *Anacamptis* (*Orchis*) *pyramidalis* (Fig. 2.1). He went on to remark in the first edition of his book on orchid pollination (Darwin 1862, p. 46) that "he who believes this doctrine must rank very low the instinctive knowledge of many kinds of moths." In the second edition (Darwin 1877, p. 36) he was emboldened enough to generalize this

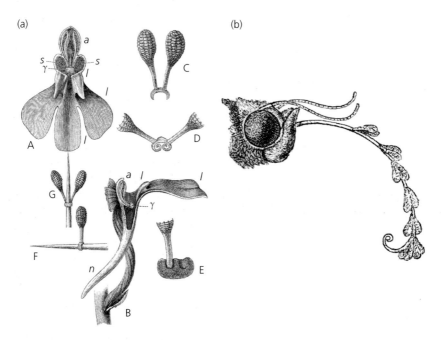

Figure 2.1 Drawings from the first edition of Darwin's book on orchid pollination showing (a) the flower structure of *Anacamptis* (= *Orchis*) *pyramidalis*, including the reconfiguration of freshly withdrawn pollinaria, and (b) a noctuid moth carrying pollinaria of this orchid. Darwin mistakenly believed that the presence of multiple pollinaria on the moth's proboscis could only be explained by the presence of nectar in the spurs. From Darwin (1862, pp. 22, 37).

statement to "he who believes in Sprengel's doctrine must rank the sense of instinctive knowledge of many kinds of insects, even bees, very low in scale."

Darwin was considerably troubled by the problem of empty orchid flowers, and after observations lasting 23 consecutive days, he was forced to concede that the floral spurs of *Anacamptis* (as *Orchis*) *morio* were "dry." Yet, after finding that fluid could be exuded from the cut ends of the spurs, he mistakenly concluded that the nectar must instead be hidden within the plant tissue and accessed only when the epidermis is pierced by the proboscides of insects. Darwin managed to convince Hermann Müller, then the most influential pollination biologist in Europe, that insects obtained their reward from *Orchis* flowers in this manner.

Darwin even attempted an experiment to test whether flowers of *A. pyramidalis* contain nectar. He removed the spurs from six flowers of this species and predicted that this would discourage pollinator visits. He seemed jubilant when just three of these flowers were visited, as opposed to 13 out of 15 intact control flowers. However, as pointed out by Ackerman (1986), not only was this a flawed experiment (removal of the spur also affects the morphological fit between insect and flower), but was also hopelessly under-replicated and therefore statistically non-significant. Indeed, we now know that *A. pyramidalis* is a deceptive species and that Darwin, for all his brilliance, was no statistician.

The prevalence of rewardlessness among plants

The most thorough survey of the extent of floral deception in angiosperms was conducted by Renner (2006). She identified the phenomenon in 32 families and estimated that several hundred genera, making up about 3.7% of angiosperm species, have deceptive flowers. Stefan Vogel (1993) had earlier estimated a figure of 6%. The greatest concentration of rewardless species is found in the Orchidaceae, accounting for 30–40% of the approximately 27 000 species in the family. Renner estimated that a further 1000 deceptive species are found in other plant families, including the Araceae, Aristolochiaceae, Begoniaceae, and Apocynaceae.

It must be emphasized that these are only very rough estimates, since rewardlessness is a labile trait and often varies among species in a genus (Johnson et al. 2013). It can even vary between male and female flowers and, more rarely, among populations and even individuals of hermaphrodite species (Brown and Kodric-Brown 1979; Sazima and Sazima 1989). In addition, since rewardlessness cannot easily be assessed from herbarium specimens, available information on the prevalence of rewards in plants is restricted to those species that have been specifically examined for this trait in the field.

Rewardlessness also occurs in insect-pollinated gymnosperms. In most cycads, for example, male cone tissue is used as a brood site for insects that vector pollen, and the female cones—which are effectively rewardless—are pollinated when these insects visit them by mistake (Donaldson 1997).

Phylogenetic evidence for transitions between reward and deception

There is now ample evidence for repeated transitions between deceptive and rewarding pollination systems in plants. In many orchid clades deception is the ancestral state and nectar has evolved secondarily (Cozzolino and Widmer 2005). In the African orchid genus *Disa*, for example, deception is ancestral and there have been at least nine transitions to rewards and one reversal from rewards back to deception (Johnson et al. 2013). Similar patterns emerge from studies of terrestrial orchids in Europe and Australia (Inda et al. 2012), although in those lineages transitions between deception and reward are mostly only apparent at or above the genus level. For the European orchidoid orchids, Smithson (2009) calculated that there were probably eight transitions from ancestral deception to nectar production. In their phylogenetic study of the Australian diurid orchids, Weston et al. (2014) concluded that deception was ancestral and that there have been nine to twelve transitions to nectar production. There is also evidence for a shift from deception to nectar production in the orchid subfamily Vanilloideae (Pansarin et al. 2012), and multiple origins of nectar production and at least three transitions from deception to the production of fragrance rewards in the huge orchid subfamily Epidendroideae (Ramirez et al. 2011).

This lability in nectar production runs counter to an earlier suggestion by Gill (1989) that the evolution of rewards in orchids must be constrained by an absence of suitable mutations. The multiple independent origins of nectar production in different clades in *Disa* is also evidenced by the evolution of nectar secretion in non-homologous parts (such as sepal spurs and petals) in different lineages, as well as the wide variety of epidermal structures associated with nectar secretion (Hobbhahn et al. 2013). Lability in nectar production is all the more remarkable given the recent evidence that secretion of nectar is a complex biochemical process (Brandenburg et al. 2009; Heil 2011). Vascular traces do not reach the epidermis in most nectaries and nectar secretion is thus not a simple case of channeling phloem to the outside of the plant (Heil 2011). The mechanisms involved in the multiple independent origins of nectar secretion in lineages that had been deceptive for tens of millions of years will be understood only once the genetic architecture of nectar secretion is fully revealed. One clue as to how nectar is secreted from different floral parts (and even from vegetative parts) in different lineages is that genes that specify nectaries appear to be independent of the ABC genes that regulate floral organs (Brandenburg et al. 2009).

It seems highly likely that in orchids rewardlessness evolved before mimicry. Evidence from studies of European, Australian, and African terrestrial orchids suggests that Batesian food-source and sexual mimicry were derived from generalized food deception (Johnson et al. 1998; Inda et al. 2012; Weston et al. 2014). Nevertheless, there are also instances of mimicry evolving in lineages that produce food rewards (Bänziger 1996b), underscoring our point, also discussed in Chapter 3, that deception and mimicry are not synonymous.

The evolution of empty flowers: cost-based models

During a typical foraging bout, animal visitors to flowers often encounter flowers that are rewardless, or nearly so, because the rewards have been depleted by previous visitors. Animals adjust their foraging by leaving patches with relatively low rewards or high variability in reward availability according to decision rules, such as those proposed in the marginal value theorem (Charnov 1976). Importantly, foragers will not necessarily shift their foraging to another species simply on account of having encountered some transiently rewardless flowers (Chittka et al. 1997). This natural reality opens the possibility that rewardlessness could evolve as a cheating strategy. Such a possibility is of considerable interest, as it may provide the initial circumstance whereby a shift from reward to rewardlessness might occur within a lineage.

The most influential cost-based model was proposed by Bell (1986). He outlined a coevolutionary scenario whereby highly discriminating pollinators select for consistent nectar production by plants, but since these pollinators would waste time on discrimination, selection favors lower levels of discrimination, which, in turn, favors the evolution of plant cheats. A very similar idea, namely that selection for discrimination ability depends on the costs of being deceived, was proposed by Lehtonen and Whitehead (2014) for the evolutionary dynamics involving plants that mimic female insects and are pollinated by male insects (see Chapter 5 on sexual mimicry for further details). However, the costs of being deceived have proved difficult to measure, particularly in the case of food-based deception. There has been more success measuring the cost of deception for mate-seeking animals in terms of lost mating opportunities (through the ability and time taken to locate genuine females) when they are attracted to sexually deceptive flowers (Wong and Schiestl 2002; de Jager and Ellis 2014).

Using game theory, Bell (1986) considered the evolution of empty flowers for a single-species system in terms of entire plants being either rewarding or cheating and in terms of plants having a certain proportion of rewardless flowers. Bell assumed that the main advantage to a plant of having empty flowers was the saving of physiological resources that could be allocated to seed production. He assumed that it would pay pollinators (in terms of energy gain) to be undiscriminating when a small proportion of flowers are empty and to be discriminating when a higher proportion are empty. This would be influenced by the discrimination time

(D) relative to the handling time (H) when feeding on flowers. Larger values of D would select for a greater proportion of cheating flowers. Bell's ideas were later echoed by Dawkins and Guilford (1991), who argued that the correlations between the frequency of cheating and discrimination time also apply more generally to interactions among animals that use signals for fighting and mating, and make it likely that honest signaling is not as pervasive as predicted in earlier models.

In simulations, Bell found that the proportion of empty flowers on a plant would be roughly equal to D/H. There is some empirical support for this association from a study of *Cerinthe major* (Boraginaceae), but the prediction from D/H was met for the proportion of flowers that produced low amounts of nectar rather than the proportion of entirely empty flowers (Gilbert et al. 1991). In addition, those producing low amounts of nectar may have simply been old flowers in the sample. In support of Bell's model, Thakar et al. (2003) also reported high proportions (up to 68%) of empty flowers on inflorescences of 24 different plant species that had been bagged to exclude pollinators before nectar sampling. We feel that caution should be exercised when interpreting these findings—firstly, because it is often very difficult to measure nectar accurately in flowers, and secondly, because nectar production in plants is very sensitive to flower age and environmental conditions, such as water availability (Zimmerman 1983).

Bell (1986) did not elaborate on exactly how pollinators would discriminate against cheats, but did suggest that pollinators would find it harder to discriminate against cheats when flowers conceal nectar (e.g., in deep tubes or spurs), and so flowers with such designs should be more likely to be cheaters. Smithson and Gigord (2003) further suggested that avoidance by pollinators of deceptive morphs by learning their location or by departing from patches where such plants are aggregated would also represent a form of remote discrimination that is costly in terms of time. If animals cannot discriminate in any way between rewarding and cheating flowers of a species, then cheating by plants should always be favored, unless animals can shift between species—which is of course a common phenomenon in natural communities (Chittka et al. 1997; Cartar 2004).

Bell noted that in such a case wholly cheating species would need to imitate the signals of sympatric rewarding species to avoid being neglected by pollinators. This leads to the theoretical basis of Batesian food-source mimicry, which is discussed in Chapter 4.

As a test of how Bell's idea might apply in a two-species system, Smithson and Gigord (2003) simulated the consequences of different pollinator foraging strategies when one species had varying frequencies of rewardless and rewarding flowers while the other was consistently rewarding. They found that pollinators could come close to achieving the energy gain obtainable by remote discrimination simply by foraging randomly and departing rapidly from non-rewarding models. Switching to an alternative species was only optimal for high frequencies of non-rewarding models. They then performed experiments in which bumblebees were presented with varying proportions of empty versus rewarding purple model flowers and consistently rewarding yellow model flowers making up 33% of the models. Bees shifted strongly to yellow flowers only when the frequency of non-rewarding purple models approached 90–100%. They never completely avoided purple models, but showed a small but significant degree of remote discrimination against empty purple models in favor of rewarding ones, which increased with the frequency of empty models. Smithson and Gigord's model and experiments allowed pollinators not only to use remote discrimination, but also to shift between species and to reduce the time spent on empty inflorescences. As such it represents a significant improvement on Bell's original single-species model.

Bell (1986) dismissed the "pure" strategy—plants having all empty flowers—as unrealistic because an "insect could quickly learn to leave a plant after encountering a single empty flower." Yet the pure strategy is the one that is found in thousands of plant species, while Bell's alternative "mixed" strategy of plants having a certain proportion of empty flowers remains controversial and poorly documented in nature.

We now consider two of the central tenets of Bell's model: that nectar has a significant physiological cost; and whether pollinators can remotely detect empty flowers in a rewarding species.

The costs of nectar

Estimates of the cost of nectar production vary widely, ranging from as much as 33% of the cost of producing the long-lived flowers of *Asclepias* species (Pleasants and Chaplin 1983; Southwick 1984) to just 3.3% of the cost of the short-lived flowers of *Pontederia cordata* (Harder and Barrett 1992). A meta-analysis in the latter study suggested that the estimate for *P. cordata* may be close to the true median for plants. Williams et al. (1985) even showed that some flowers can support their own nectar production through photosynthesis, making it unlikely that the cost of nectar is a major constraint on flower production per se. Indeed, Golubov et al. (2004) were unable to detect any differences in growth and fecundity between rewardless and rewarding morphs of the honey mesquite *Prosopis glandulosa* var. *torreyana*.

One study has provided evidence that nectar can be costly for plant reproduction: Pyke (1991) discovered that flowers of *Blandfordia nobilis* that had been induced to produce more nectar by regular nectar removal had a reduced capacity for seed production following hand pollination, relative to control flowers from which nectar had not been removed. This suggests that nectar may have significant costs in terms of a currency (seeds) that is important for fitness, particularly in plants that are not pollen limited and secrete relatively large amounts of nectar, as is the case for the bird-pollinated *B. nobilis*. These experiments also suggest that nectar costs are not fixed and that plants that receive many visits (requiring that nectar be re-secreted) may thus pay a higher gross cost in terms of nectar production.

Another clue that production of nectar may involve significant costs for plants is that many species actively reabsorb nectar (Nepi and Stpiczynska 2007). Sugars from reabsorbed nectar have been traced to various plant parts, including fruits. Nectar reabsorption in some orchids seems to be triggered by pollination (Koopowitz and Marchant 1998; Luyt and Johnson 2002), which makes sense given that the sexual function of an orchid flower in terms of pollen deposition and removal is often accomplished during a single visit from a pollinator. The potential fitness benefits of nectar reabsorption were demonstrated in a study of the epiphytic orchid *Mystacidium venosum* (Luyt and Johnson 2002). In that study, plants prevented from reabsorbing nectar after pollination produced fruits with a lower percentage of viable seeds than those that were able to reabsorb nectar.

In general, we feel that available data are both too few and too variable to support firm pronouncements concerning the cost of nectar production and its role in the evolution of deception. The cost of nectar (in relation to the diminishing returns on this investment for pollinator visitation) may explain why plants do not produce very large amounts of nectar, but it does not explain why many plants are completely deceptive. We also know little about the costs of other nectar constituents, such as amino acids, secondary metabolites, and water. Costs are unlikely to be so prohibitive that they exclude plants from producing even very small amounts of nectar. Furthermore, many rewardless orchids produce copious extra-floral nectar, suggesting that the absence of floral nectar is not caused by resource limitation.

Can pollinators remotely detect empty flowers in a rewarding species?

Bell (1986) considered nectar concealment as a strategy whereby plants could limit the detection of empty flowers by pollinators. There is mixed evidence about the abilities of pollinators to assess the availability of nectar in flowers without actually landing on or probing them. The success of deceptive orchids means that, in general, insects—even those with a good memory capacity such as bees—cannot easily detect and thus avoid empty flowers. Yet there is good empirical evidence that pollinators are less likely to probe flowers with low nectar rewards, even when nectar is concealed (Marden 1984). The most obvious mechanisms are that pollinators use visual cues associated with flower aging or previous visitation, scent—either "footprints" left by previous visitors (Schmitt and Bertsch 1990) or the scent of the nectar itself—or possibly even humidity (von Arx et al. 2012) as measures of nectar availability (Heinrich 1979; Marden 1984). In one experiment (Fig. 2.2), *Osmia* bees (Megachilidae) prevented from smelling

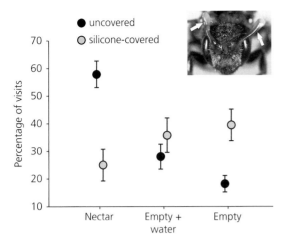

Figure 2.2 *Osmia* bees with uncovered antenna prefer *Penstemon caesius* flowers that contain nectar. This preference disappears when the bees' antennae are covered with silicone (arrows on inserted image) to prevent olfaction. Adapted from Alarcon and Howell (2007). Photo © Anna Howell.

nectar due to the application of a coating of silicone over their antennae could not discriminate between empty and nectar-rich flowers of *Penstemon caesius* (Scrophulariaceae), whereas bees with unmanipulated antennae visited nectar-rewarding flowers twice as often as empty ones (Howell and Alarcon 2007), indicating that their discrimination against empty flowers is based on sensory evidence.

Plants that deploy scented nectar are less likely than those with unscented nectar to cheat pollinators into visiting empty flowers; the existence of plants with scented nectar thus contradicts Bell's idea that selection should favor nectar concealment and cheating. Nectar is sometimes scented and distinctive from the rest of the flower, suggesting that pollinators such as bees and moths could avoid empty flowers using smell. Raguso (2004) argued that honest signaling via scented nectar would benefit plants in terms of being more attractive to pollinators and also by directing visitors to receptive, unvisited flowers. Similar arguments have been used to explain color change in pollinated flowers (Weiss 1991). The counter-argument would be that for most plants, multiple visits would be required to remove pollen, even from visited flowers, and so even individual nectar-depleted flowers would benefit from remaining attractive to pollinators.

Pollinators could also discriminate against deceptive plants by remembering their location (Makino and Sakai 2007) and, in the case of honey bees, even communicating this information to other individuals via their dance language. Klinkhamer et al. (2001) conducted interesting experiments with groups of *Echium vulgare* varying in nectar production rate (NPR). Plants with a low NPR benefited from being closely surrounded by individuals with a high NPR, but if groups were separated by more than 6 m, plants with a low NPR received fewer approaches. Subsequent experiments with potted *E. vulgare* plants confirmed that the approach rates depended on the NPR in sparse populations but not in dense populations (Klinkhamer and van der Lugt 2004). This context dependence of selection on NPR could explain why rewards are so variable among plants. Many deceptive orchids occur at very low densities and would be prone to being neglected due to spatial learning by pollinators. However, this could be mitigated if these isolated orchid plants require very few visits for pollination and make use of pollinators that move constantly and are not local residents.

Nectar-deployment models based on mating patterns

Models for the evolution of empty flowers that emphasize the physiological costs of nectar (Bell 1986; Thakar et al. 2003) could be criticized for overlooking the importance of plant mating patterns. Some studies suggest that the evolution of deceptive flowers can be explained even if the cost of nectar is negligible. Harder and Barrett (1992), for example, argued that the main selective advantage for small, as opposed to large, amounts of nectar in flowers is related to the proportion of pollen removed by a pollinator that is lost during transport. Limited pollen removal during each of numerous short flower visits will result in lower transport losses and therefore more successful pollen dispersal than the removal of abundant pollen during a single, long visit to a flower.

Bell (1986) and Pyke (1991) considered the limitation on plant investment in nectar to be its cost. However, plants that invest in attractive traits such as nectar also face a paradox, because their increased attractiveness can also lead to undesirable mating patterns

(Klinkhamer and de Jong 1993). Bell considered a pollinator's departure from a plant after encountering empty flowers to be disadvantageous for the plant, but such behavior may actually enhance cross-pollination and reduce the deleterious consequences of geitonogamous pollination (Johnson and Nilsson 1999; Johnson et al. 2004; Jersáková and Johnson 2006). The paradox that increased attractiveness of plants can reduce the quality of mating has been considered mainly with respect to the number of flowers displayed by a plant (Harder and Barrett 1995), but it applies equally to the deployment of nectar.

The patterns of pollen transfer that influence plant mating can be modeled as a series of exponential decay curves. For each flower visited in a sequence, a certain proportion of pollen will be picked up and a certain proportion of this will be deposited on flowers visited subsequently. As more flowers on a plant are visited, a greater fraction of a plant's exportable pollen will be deposited on its own stigmas, a phenomenon known as "pollen discounting" that reduces male fitness (Harder and Barrett 1995). It can also compromise female fitness through "ovule discounting" if self-pollination reduces the number of ovules available for production of outcrossed, and therefore higher-quality, progeny (Barrett et al. 1996).

Models by Bailey et al. (2007) incorporated mating systems and considered the effect of the proportion of empty flowers in inflorescences on geitonogamy and the costs of self-pollination (which they refer to as selfing). These models differ from Bell's model in that they ignore the cost of nectar and show that empty flowers could evolve even without consideration of the cost of nectar. In both analytical and simulation versions of the mating-system model, inflorescences with a high proportion of empty flowers were found to be optimal when pollinators are abundant and costs of self-pollination for male mating opportunities are high because of inbreeding depression or self-incompatibility and pollen discounting.

Supplementation of nectar in rewarding flowers

There is a long tradition of studies investigating how pollinator behavior and plant fecundity are influenced by the amount of nectar available in flowers. Researchers have used natural variation in nectar production, artificially supplemented nectar in flowers, and have even stimulated nectar production by watering plants (Zimmerman 1983). Larger amounts of nectar generally increase the number of approaches by pollinators, the number of flowers probed by pollinators, and the duration of each probe (Thomson and Plowright 1980; Waddington 1981; Real and Rathcke 1991; Hodges 1995). These behaviors, in turn, increase pollen removal and deposition on flowers (Galen and Plowright 1985), and in some cases increase fruit production. However, increases in nectar generally provide diminishing fitness returns (Burd 1995): doubling nectar seldom doubles pollen deposition and removal. Diminishing returns, considered with the cost of additional nectar as well as the risk that increased nectar increases the incidence of geitonogamy, may explain why most plants produce relatively small amounts of nectar.

What can we learn from studies of nectar robbing?

Numerous studies have investigated the impact of nectar robbing on plant fitness; these often involve the experimental removal of nectar from flowers. Such studies are of interest to those interested in a different problem—the success of rewardless mutants in plant populations. We should clarify that removal of nectar either by a floral larcenist or by a researcher using a pipette (Irwin and Brody 1998) will not result in a completely empty flower, so these manipulations are likely to underestimate the actual effects on animal behavior of a mutation for rewardlessness. These studies also tend to involve nectar depletion in a subset of flowers, whereas a mutation for rewardlessness would affect entire plants. Furthermore, a rewardless mutant would compete with rewarding conspecifics in the same population, whereas floral larceny is likely to affect entire populations. Despite these caveats, studies of nectar robbing should shed light on some of the problems examined in this book.

Nectar robbing tends to affect pollinator behavior as predicted by the marginal value theorem

(Charnov 1976)—animals visit fewer flowers and probe flowers for a shorter time when flowers have been robbed (Irwin and Brody 1998). These behaviors can also reduce geitonogamy, but a study of the effects of nectar robbing by bumblebees on the pollination of *Ipomopsis aggregata* by hummingbirds indicated stronger negative effects of nectar robbing on female fecundity than any positive effects through reduced geitonogamy (Irwin 2003). In general, nectar robbing tends to cause a small but significant decrease in female fecundity of plants (Irwin et al. 2001; Irwin et al. 2010). Only a few studies have included measurements of the effects of nectar robbing on male function; these have found either no effects or negative effects overall (Irwin et al. 2010). Of interest, given the spatial genetic structure of plant populations, is that at least two studies have found that nectar robbing increases pollinator flight distances (Zimmerman and Cook 1985; Maloof 2001). A handful of studies have shown some positive effects of nectar robbing, but these are usually due to pollination activity by the robber rather than the effect of nectar removal per se. These studies are consistent with our general conclusion that for most plants nectar is important for reproduction and that selection does not generally favor rewardlessness. However, it should also be noted that many studies have not detected significant negative effects of robbing on fitness, which may relate to the point made earlier in this chapter that most flower-feeding insects are not deterred by encountering a certain percentage of empty flowers during their foraging bouts, particularly in heavily visited populations with a high percentage of recently visited flowers that are temporarily empty.

The puzzle of floral deception in orchids

The commonness of floral deception in orchids is a major evolutionary puzzle, because there is now overwhelming evidence that deceptive species are, on average, much less fecund than rewarding species. Neiland and Wilcock (1998) showed that the percentage of flowers that set fruit in orchids varies among geographic regions, but that fruit set of deceptive species in their survey was consistently less than half that of rewarding species (for example,

12% versus 25% for tropical species, and 28% versus 63% for European species). A subsequent analysis of fruiting failure in a larger set of species (Tremblay et al. 2005) yielded very similar results (Fig. 2.3). There seems little doubt that the lower fecundity of deceptive species is due to pollination limitation resulting from a failure to attract visitors, as studies have found that the frequency of stigmas with pollen deposited on them is much lower in deceptive than in rewarding species (Johnson and Bond 1997; Scopece et al. 2010), and that supplemental pollination is more likely to boost fecundity in deceptive species than it is in rewarding species (Johnson and Bond 1997; Tremblay et al. 2005).

In an influential essay, Douglas Gill (1989) concluded that floral deception in orchids cannot be an evolutionarily stable strategy (ESS). Gill came to this conclusion after finding that only 3% of the 1031 flowers of the food-deceptive orchid *Cypripedium acaule* that he had tracked in the forests of Virginia had actually set fruit. Fruiting failure in this species, as with most orchids, is clearly attributable to pollen limitation as hand-pollinated flowers invariably set fruit. Gill wondered why selection didn't favor selfing or reward production, as mutants with these attributes would be almost guaranteed to be "instantly successful." (An interesting postscript that Gill communicated to us is that after tracking more than 6000 individual marked plants of this species over a period of 39 years, he has discovered that

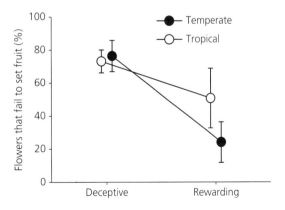

Figure 2.3 The incidence of fruiting failure is higher in deceptive than in rewarding orchids. The lowest incidence of fruiting failure is found among rewarding orchids in temperate regions. Adapted from Tremblay et al. (2005).

fruit set improves markedly after the orchid's forest habitat is burnt or defoliated by gypsy moths. According to Gill, this is probably because such opening up of the forest favors the growth of co-flowering rewarding ericaceous shrubs that act as magnet species to attract the bumblebees that pollinate the orchid.)

Mutations for nectar production in deceptive orchids have been simulated by artificial addition of nectar to flowers of various species in situ, but this has not resulted in the dramatic increases in fecundity predicted by Gill. One-off addition of nectar to spurs in a Swedish population of *Anacamptis morio* increased pollination success at only one of two sites and did not affect overall fruit set (Johnson and Nilsson 1999). Even the more biologically realistic treatment of daily addition of nectar to flowers had no effect on fruit set in an experiment conducted on this species in a population in France (Smithson 2002). In a study on *Dactylorhiza sambucina* in the Czech Republic, daily addition of nectar to the floral spurs increased pollination success in both color morphs of this species by about 30%, but increased the removal of pollinaria in the yellow morph only (Jersáková et al. 2008). Daily addition of nectar to the flowers of yet another European orchid, *Himmantoglossum* (*Barlia*) *robertiana*, had no influence on fruit set (Smithson and Gigord 2001).

In summary, the evidence from interspecies comparisons clearly shows that nectar production tends to double the rate of fruit production, whereas experiments using nectar addition show more modest effects of nectar on fruit set. This latter pattern is likely to be due to the methodological problems associated with nectar addition and the fact that in these experiments nectar is usually added to only a small fraction of the population of the rewardless species, and is thus unlikely to influence the attraction of pollinators to the population as a whole.

These caveats aside, fruit production (or even pollination rate) is not necessarily an accurate indication of the potential fitness benefits of a trait. To underline this point, it is well known that autonomous self-pollination causes high fruit production in orchids, yet the orchid family is not dominated by autonomous selfers. To solve the problem of deception in orchids we must therefore take a more nuanced view. One nuance is, of course, the cost

of nectar, and deceptive plants may have more resources to allocate to other life-history functions. However, the cost of nectar should be no different for orchids than for other plants, meaning that explanations for the maintenance of rewardlessness in orchids need to go beyond arguments about the cost of nectar. Another issue is that fruit set may only reflect the quantity of pollination, whereas fitness may differ according to the quality of pollination, specifically in terms of its influence on outcrossing and the efficiency of pollen export.

Could there perhaps be something about the patterns of pollen transfer in orchids that would explain why so many of these plants are deceptive? This leads us to consider two hypotheses for the evolution of rewardlessness in orchids: the pollen transport efficiency hypothesis and the cross-pollination hypothesis.

The pollen transport efficiency hypothesis

Harder (2000) proposed that deception is widespread among orchids because the "packaging" of their pollen in pollinaria allows total pollen export to be maximized while requiring fewer pollinators than are needed for species with granular pollen. His argument focused on the decelerating relation of pollen export to removal by individual pollinators experienced by species with granular pollen. He attributed this to transport losses arising from overlayering of pollen and grooming by pollinators. Theoretical models (Harder and Thomson 1989; Harder and Wilson 1994) indicate that such diminishing returns promote restricted removal of pollen by individual pollinators and the attraction of many pollinators to maximize total pollen export.

In contrast to granular pollen, orchid pollen is less susceptible to transport loss because it is usually glued firmly to the body of the pollinator by means of a sticky viscidium (Fig. 2.4). A meta-analysis of pollen fates in angiosperms has shown that pollen transport losses are indeed much lower in orchids than in plants with granular pollen. For example, typically less than 1% of removed granular pollen reaches conspecific stigmas, whereas the equivalent figure for orchids is 10–40%, depending on the type of pollinarium (Harder and Johnson 2008). This is reflected in the much lower pollen–ovule ratios of

Figure 2.4 Almost all orchid species package their pollen into pollinaria that are glued to the body of the pollinator, thereby increasing the efficiency of pollen transfer between flowers. In this image a horse fly (*Philoliche rostrata*) is carrying pollinaria of the food-deceptive South African orchid *Disa harveyana.* (Photo © Steve Johnson).

most orchids compared with plants that produce granular pollen. If reduced transport loss relaxes diminishing export returns from increased pollen removal, then removal by individual pollinators need not be restricted as much in order to maximize total export, and fewer pollinators need to be attracted. Under such conditions, deception may become feasible.

The packaging of orchid pollen into pollinaria means that the sexual function of flowers in terms of pollen export and deposition can often be achieved by a single pollinator visit. In contrast, plants with granular pollen usually require numerous visits to dispense pollen from a flower. However, while it is true in theory that the flowers of most orchids can export their full complement of pollen in a single visit, in reality this is not always the case. In *Anacamptis morio*, for example, the probability of pollinaria being removed from a flower probed by a pollinator was just 20% in a Swedish population (Johnson et al. 2004) and 50% in a French population (Smithson 2002). In these studies, addition of nectar either doubled the probability of removal of pollinaria (Johnson et al. 2004) or had no overall effect (Smithson 2002). In *Disa pulchra*, a deceptive South African orchid pollinated by long-proboscid flies, commonly only one of the two pollinaria is removed when a flower is probed, as is the case

for many other orchids (Harder 2000). On average, only 30% of the pollen complement in flowers of *D. pulchra* is removed per visit, and this increases two-fold when nectar is added to flowers (Jersáková and Johnson 2006). By contrast, daily addition of nectar to flowers of the European orchid *Himmantoglossum* (*Barlia*) *robertiana* actually decreased the rate of removal of pollinaria (Smithson and Gigord 2001). This latter result is especially puzzling, and may reflect either a change in the probing behavior of bees or a negative effect of nectar on the adhesive power of the viscidia.

Apart from the fact that in orchids pollen removal is not always assured during a pollinator visit, they also tend to have much higher failure of pollen removal over the lifespan of the flower. Harder (2000) showed that about 49% of orchid pollen typically remains in flowers, whereas the equivalent median figure for plants with granular pollen is just 7%. Failure of pollen removal in orchids is probably due to the rarity of pollinator visits, which in turn is correlated with a lack of rewards. Indeed, Harder (2000) found that pollen-removal failure is almost twice as high in deceptive orchid species than in rewarding species (64% versus 35%), which is consistent with the empirical results of most nectar-addition experiments involving orchids that are discussed elsewhere in this chapter.

We do not doubt the validity of Harder's general argument about deception being a more viable option for orchids on account of their increased pollen transport efficiency. However, the fact that many visits of pollinators to orchids do not result in pollen removal and that much orchid pollen is never removed from flowers further adds to the puzzle of the possible reasons for the evolutionary maintenance of deception. Other complementary explanations for floral deception will now be considered.

The cross-pollination hypothesis

The cross-pollination hypothesis for floral deception can be traced to Dafni and Ivri (1979) and Dressler (1981), and received empirical support from pollen-labeling studies conducted by Peakall and Beattie (1996). Johnson and Nilsson (1999) elaborated on this hypothesis to incorporate the theoretical ideas developed by Klinkhamer and de Jong

(1993) about trade-offs between floral attractiveness and pollination quality. The hypothesis essentially proposes that deception enhances fitness because it results in short visitation sequences on plants, thus reducing geitonogamy and enhancing cross-pollination. Klinkhamer and de Jong (1993, p. 180) expressed this idea as follows: "To maximize pollen export and to avoid geitonogamy a plant should ideally receive an infinite number of approaches by pollinators that visit a single flower during each approach. This puts the plant in a dilemma: almost all features that lead to increased attractiveness to pollinators also lead to longer flower visitation sequences."

The cross-pollination hypothesis applies to plants generally, but is particularly applicable to orchids for the following reasons: (1) Orchids, particularly those with solid pollinaria, would be expected to have limited pollen carryover and thus be prone to severe geitonogamy if animals visit many flowers on a plant. (2) The vast majority of orchids are self-compatible and therefore vulnerable to production of lower-quality progeny if self-pollinated. (3) Studies of embryo development in orchids suggest that they may experience very severe pre-dispersal inbreeding depression following self-pollination or bi-parental inbreeding (Tremblay et al. 2005).

Earlier in this chapter we showed that augmentation of nectar in already-rewarding flowers causes pollinators to probe more flowers per plant, probe individual flowers for longer, and move between closer plants. Are similar patterns observable when nectar is added to rewardless orchid flowers, and does this decrease the efficiency of cross-pollination?

Evidence overwhelmingly supports the prediction that a mutation for nectar production in deceptive orchids would increase the number of flowers that are probed per plant by individual pollinators (Fig. 2.5). Thus flower visitors interpret the flowers of deceptive orchids as non-rewarding, contrary to speculative suggestions that the papillate texture of the inner surface of the spurs of some rewardless orchids may give the impression that plants are potentially rewarding (Bell et al. 2009).

In studies of European deceptive orchids pollinated by bumblebees and South African orchids pollinated by long-proboscid flies, addition of nectar

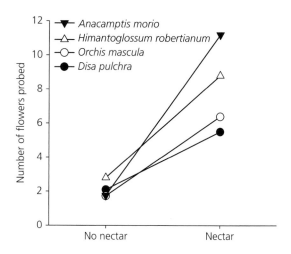

Figure 2.5 Experimental addition of artificial nectar to the flowers of food-deceptive orchids usually markedly increases the number of flowers probed by individual pollinators. Adapted from Johnson and Nilsson (1999), Smithson and Gigord (2001), and Jersakova and Johnson (2006).

caused roughly five-fold increases in the number of flowers probed, from one to two flowers per plant when spurs were empty, to six to twelve flowers per plant when spurs contained nectar (Johnson and Nilsson 1999; Smithson and Gigord 2001; Smithson 2002; Johnson et al. 2004; Jersáková and Johnson 2006) (Fig. 2.5). The effects of nectar addition on probing time per flower were more variable, with some studies showing no effect (Smithson and Gigord 2001), some showing a two-fold increase (Johnson and Nilsson 1999; Johnson et al. 2004), and one showing a six-fold increase (Jersáková and Johnson 2006). Pollinator flight distances following the addition of nectar to rewardless flowers have been examined in only one published study: Jersáková and Johnson (2006) found that addition of nectar to the spurs of the deceptive orchid *D. pulchra* halved the frequency of flights further than 40 cm, from 80% to 40%. This behavior is likely to increase the probability of bi-parental inbreeding if neighboring plants are close relatives due to limited pollen and seed dispersal.

Pollen fates in orchids can be studied by color labeling of pollen with histochemical stains or even by attaching unique numbered tags onto pollinaria (Nilsson et al. 1992). Contrary to some claims based on limited sampling (Kropf and Renner 2008), the available data from species comparisons suggest that the frequency of self-pollination is generally higher in rewarding than in deceptive species. As a general rule of thumb, self-pollination makes up more than 30% of all pollen transfers in rewarding species and less than 20% of pollen transfers in deceptive species (Jersáková and Johnson 2006; Jersáková et al. 2006 and references therein; Johnson et al. 2009).

The cross-pollination hypothesis can also be tested by adding nectar to flowers of a deceptive species and comparing the fates of pollen removed from these flowers with those of pollen from control deceptive flowers. Johnson et al. (2004) used histochemical stains to label pollen of nectar-supplemented plants of the orchid *A. morio*, so that self- and cross-pollen could be distinguished on stigmas. They found that self-pollination rarely occurs when inflorescences are deceptive and that a mutation for nectar production would result in a rate of self-pollination of about 40%. In a similar study of a non-rewarding South African orchid pollinated by long-proboscid flies, Jersáková and Johnson (2006) concluded that a mutation for nectar production would increase the rate of self-pollination from 27% to 70% if all flies that arrived at a plant were already carrying pollinaria. In reality, only half of the flies in their study arrived with pollinaria and so they estimated that the natural rate of pollinator-mediated self-pollination could be as high as 90% for rewarding plants. The effect of long visitation sequences on the proportion of flowers on rewarding plants that were self-pollinated was clearly evident from their results (Fig. 2.6).

The relation of the incidence of geitonogamy to the number of flowers probed in orchids depends on whether the species' pollinaria reconfigure after removal to facilitate pollen deposition. Reconfiguration of pollinaria extends the interval between pollen removal and deposition on stigmas (Darwin 1862; Peter and Johnson 2006). The most common mechanism is for the strap-like caudicle or stipe of freshly removed pollinaria to slowly bend so that the pollinium (pollen mass) can contact stigmas (Fig. 2.1). Reconfiguration of pollinaria limits within-flower and geitonogamous self-pollination in orchids (Johnson and Nilsson 1999; Johnson et al. 2004), so selection on floral rewards is

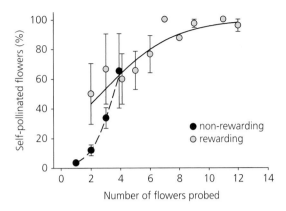

Figure 2.6 The percentage of flowers on a plant that experience pollinator-mediated self-pollination generally increases with the number of flowers that are probed by the pollinator. These data for the orchid *Disa pulchra* show that addition of a nectar reward increases the number of flowers probed and, in turn, the incidence of self-pollination. Adapted from Jersáková and Johnson (2006).

correlated with selection on pollinarium reconfiguration. The duration of pollinarium reconfiguration is finely adapted to the length of time that pollinators typically stay on a flower (Peter and Johnson 2006). Thus slower pollinarium reconfiguration should evolve in nectar-rewarding orchids, because pollinators spend more time on inflorescences and on individual flowers.

The evolution of floral deception (and its maintenance by selection) is a complex problem because deception has contrasting effects on different components of reproductive success. On one hand deception decreases overall plant attractiveness and reduces fruit production, but on the other hand it increases the efficiency of cross-pollination and hence can improve seed quality and pollen export. This is a special case of the well-known trade-off between the quality and quantity of pollination. Given the multiple transitions between deception and reward in orchids, it seems reasonable to infer that the outcome of a mutation for rewards in orchids will be highly context dependent. The most important variable affecting transitions between reward and deception is likely to be pollinator availability. Put simply, rewards (promoting pollination quantity) should be selected when pollinators are rare, whereas deception (enhancing pollination quality) should be favored when pollinators are common.

Of course, when pollinators are vanishingly rare no amount of reward in flowers will solve the problem faced by plants. In such a case, selection should favor autonomous self-pollination, the strategy yielding the highest quantity and reliability, though also the lowest quality, of pollination.

The genetic load paradox

Highly outcrossed species generally carry high levels of genetic load because deleterious recessive mutations would seldom be expressed due to high heterozygosity (Husband and Schemske 1996). Conversely, selfing species are expected to carry less genetic load, because consistently high homozygosity exposes deleterious recessive alleles to selection, purging them from populations (Lande and Schemske 1985; Charlesworth and Charlesworth 1987). These theoretical predictions are supported by empirical evidence for negative correlations between selfing rates and the magnitude of inbreeding depression (Charlesworth and Charlesworth 1987; Husband and Schemske 1996). Therefore, if deceptive orchids routinely outcross more than rewarding species do, deceptive species should carry more genetic load and suffer from more inbreeding depression when selfed.

The earliest stage of inbreeding depression should occur during embryo development. Many studies have reported on the percentages of seeds with viable embryos for self-compatible deceptive and rewarding orchid species; meta-analyses of these data show that orchids frequently experience predispersal inbreeding depression (selfing typically leads to 20% fewer seeds with viable embryos compared with crossing), but they provide no support for the prediction that deceptive species carry more genetic load than rewarding species (Tremblay et al. 2005; Jersáková et al. 2006a; Smithson 2006).

Very few studies have investigated inbreeding depression in orchids during multiple life-history stages because it is extremely difficult to track progeny of known parentage in the field. This can be done for some species in the laboratory and greenhouse, but in such cases developing plants do not experience the vicissitudes of field conditions. Smithson (2006) measured inbreeding depression during protocorm development for three species by germinating seeds in the field (achieved by placing seeds in the soil in small packets that allowed access by fungal symbionts), but this did not yield clear results. No conclusions about the later stages of inbreeding depression in deceptive orchid species can therefore yet be reached.

Deception and diversification

Through its effects on outcrossing and pollinator flight distances, and its evolutionary association with mimicry, deception may enhance rates of speciation. In turn, its effect of reducing fecundity as a result of pollen limitation may increase extinction rates. What, then, is the net effect that deception has had on diversification in plant lineages?

Cozzolino and Widmer (2005) found greater genetic differentiation between populations of rewarding orchids than between those of deceptive orchids. Assuming no differences in the age of populations or distance of seed dispersal, this suggests that there is more pollinator-mediated gene flow among populations of deceptive species, which in turn would counter the effects of genetic drift and lead to high genetic variation within populations. Cozzolino and Widmer interpreted this finding as consistent with deception promoting both outcrossing in orchids and longer pollinator flights. The mechanism whereby deception would enhance pollinator-mediated gene flow between entire populations is not yet clear, but the most probable process is that pollinators do not remain within patches of deceptive orchids and thus are more likely to move between populations. Johnson et al. (2013) found no evidence that rates of speciation, extinction, and overall diversification were affected by transitions between deception and nectar rewards. A similar conclusion was reached by Smithson (2009) in her study of transitions from deception to rewards in European, Australian, and African orchid lineages. In their broad-scale phylogenetic analysis of the orchids, Givnish et al. (2015) found that deception was associated with small increases in both speciation and extinction, with a small net effect of decreased diversification within lineages and increased diversification across lineages. Therefore, the link between deception and diversification remains elusive, despite the prevalence of deception in the Orchidaceae, the world's largest family of flowering plants.

Overview and perspectives

This chapter has been devoted almost entirely to the evolutionary circumstances associated with the loss or gain of nectar as a reward. We did not deliberately ignore other rewards that flowers offer, but rather found that very little work has been done to understand the evolutionary gains and losses of these rewards in lineages that include deceptive species. What are some of the selective factors involved in transitions between deception and non-nectar/other floral rewards such as oils and fragrances? How costly are these rewards for the plant? Do these alternative rewards promote plant attractiveness, while also increasing geitonogamous self-pollination? Data presented by Neiland and Wilcock (1998) suggest that orchids that produce fragrance or oil rewards have similar fruit set to deceptive orchids, raising difficult questions about why these rewards have evolved in deceptive lineages if they do not have a fitness benefit.

There is also much uncertainty about the actual outcrossing rates in deceptive plant species. Pollen-staining experiments suggest that deceptive species outcross extensively, but what about the actual mating patterns? Mating patterns can be estimated in epidendroid orchids if each fruit is sired by pollen from a single male, and there are distinctive patterns of inbreeding depression (Peter and Johnson 2009). Unfortunately, most orchids have tiny seeds which yield little DNA or few enzymes and are difficult to germinate. Therefore, traditional methods for calculating outcrossing rates based on allele frequencies of progeny (Brown et al. 1989) are not easily applied to orchids. However, a recent study using microsatellite markers found very high outcrossing and long-distance gene flow in a deceptive Australian orchid species (Whitehead et al. 2015).

To summarize, food rewards such as nectar contribute to the attraction of pollinators through positive associative conditioning, and also increase the duration of pollinator visits and hence the probability and intensity of pollen exchange. However, food rewards can decrease the quality of pollination (by elevating levels of geitonogamy and within-flower self-pollination), and rewardlessness may therefore evolve and be maintained by selection through its advantage for promoting cross-pollination. Rewardless flowers are generally less attractive to pollinators than rewarding flowers, but this can be offset by the evolution of floral signals that exploit the receiver bias of pollinators. In the following chapters we show that exaggerated floral signals can exploit generalized foraging behavior (Chapter 3) and mimicry can result in the cognitive misclassification of flowers by pollinators (Chapters 4–7). These signaling strategies can elevate rates of visitation in rewardless (and even in rewarding) species without compromising the quality of pollination. In addition, elaborate devices such as the traps of slipper orchids can increase the duration of pollinator visits to a single flower even in the absence of food rewards.

Finally, the maintenance of floral rewardlessness needs to be viewed in a community context. Pollinator departure decisions will depend not only on the frequency of empty flowers in one species but on the overall frequency of empty flowers encountered by pollinators in a community. Avoidance learning and departure from deceptive species will be less likely if flowers of other species are temporarily depleted of nectar because of frequent visitation (Harder et al. 2001). Pollinators may therefore struggle to distinguish between rewardless and rewarding species if both transient and permanently empty flowers are common in natural communities.

CHAPTER 3

Generalized food deception

Introduction

The majority of rewardless plant species are not specific mimics of flowers of other species; instead, they deploy a generalized set of floral signals to attract pollinators. These generalized food-deceptive (GFD) species have a number of unique evolutionary and ecological properties that are discussed in this chapter. The success of GFD plants needs to be understood in light of the fact that flower-visiting animals regularly encounter reward-depleted flowers during their foraging bouts. The evolutionary strategy of GFD species succeeds because the process whereby flower visitors learn to avoid deceptive species is sometimes slow and uncertain and depends on a number of factors. These include the relative density of a plant population and the similarity of floral signals between deceptive and rewarding species (Fig. 3.1), as well as the degree of polymorphism in the signals of GFD species. Most of the known GFD species are orchids, but the principles outlined here apply broadly to other deceptive plant-pollinator interactions and underline the relative ease by which plants can manipulate food-seeking insects.

Food rewards in flowers range from nectar and pollen to oils, resins, and fragrances (Willmer 2011). In any population, a certain proportion of flowers are likely to be depleted of rewards on account of recent visitation, age, or physiological conditions preventing nectar production, such as water stress. In plants with unisexual flowers, pollen rewards are available only from male flowers, meaning that female flowers are effectively rewardless to pollen-collecting visitors (Vogel 1978; Renner 2006). In hermaphroditic flowers, rewards may only be available during temporally separated male or female stages (Vogel 1978). Flower visitors are usually unable to directly assess the reward status of a flower from a distance, and must therefore rely on floral signals to assess its profitability indirectly. This leaves visitors vulnerable to exploitation by plants that deploy dishonest signaling in the form of attractive flowers that contain no rewards. Indeed, there is some evidence that rewardless (cheating) morphs occur together with rewarding ones in populations of some species (Sazima and Sazima 1989; López-Portillo et al. 1993). Complete fixation for lack of rewards occurs in many lineages and in up to 40% of all extant species in the mega-diverse orchid family (Nilsson 1992; Jersáková et al. 2006a; Renner 2006). Contrary to some earlier speculation (Dafni 1987), rewardlessness is the ancestral condition in many orchid lineages (Cozzolino and Widmer 2005; Inda et al. 2012; Johnson et al. 2013). This raises important questions about how rewardless lineages have been so successful in terms of proliferation of species and occupation of different ecological niches. Generalized food deception is central to this debate because it applies to the majority of rewardless species.

In this chapter we outline the principle of generalized food deception and compare this pollination system with food-source mimicry, including generalized pollen mimicry. Further we show how learning by pollinators influences trait evolution, and we review patterns of selection associated with GFD. We also show that plant population size, as well as the plant community in which GFD plants grow, has a large impact on the evolutionary trajectories of GFD plants.

Floral Mimicry. Steven D. Johnson & Florian P. Schiestl.
© Steven D. Johnson & Florian P. Schiestl 2016. Published 2016 by Oxford University Press.
DOI 10.1093/acprof:oso/9780198732693.001.0001

(a)

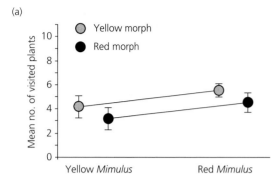

(b)

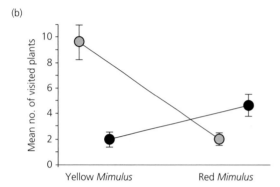

Figure 3.1 The number of bumble bee visits to yellow and red morphs of *Dactylorhiza sambucina* in relation to whether or not the bumblebees had previously visited yellow or red rewarding *Mimulus* flowers: (a) before *Mimulus* visitation; (b) after *Mimulus* visitation. The data show that bees experienced with a specific rewarding floral color prefer to visit flowers of a similar color in successive visits. Adapted from Gigord et al. (2002).

Generalized food deception and floral mimicry

Broadly, rewardless species can be classified into those employing generalized food deception and those that that use floral mimicry to deceive their pollinators. The major difference between the two systems is the lack of a specific model species in GFD plants that exploit the "general search image" of a pollinator. Plants deploying generalized food deception exploit behavior that is innate, yet also benefit facultatively from learned behavior in the signal receiver. Generalized food deception maintains its function in a variety of different communities and is not sensitive to the presence

of particular "model" species. Pollinators of GFD plants are usually generalized and will visit rewardless flowers when they are flower-naïve or during exploration of new food sources. Because of the general nature of the signals deployed by GFD plants they tend to have broad assemblages of pollinators (Nilsson 1983a; Fritz 1990; Cozzolino et al. 2005; but see Peter and Johnson 2013), in contrast to the high levels of pollination-system specificity that is characteristic of food-source mimicry (Newman et al. 2012) and sexual mimicry (Schiestl 2005). Generalized food deception is found in several plant families but is most common among orchids; more than 30% of all orchid species are thought to rely on this strategy (Jersáková et al. 2006a; Renner 2006).

Generalized food deception and floral mimicry are two different phenomena, and can usually be discriminated (Dafni 1984; Smithson and Gigord 2003; Schiestl 2005; Jersáková et al. 2006a). GFD plants lack derived traits that impart similarity to a specific model (Dafni 1984; Nilsson 1992) and tend to have showy, colorful, and weakly scented inflorescences (Dafni 1984). Mimics, on the other hand, have an adaptive resemblance to a particular model (Starrett 1993). If a trait has been favored by selection in the context of mimicry, then its evolution should be associated with the innovation of derived traits that impart a system of mimicry in that lineage (Johnson et al. 2003a). Many suggested cases of food-source mimicry fail this basic test for evolutionary novelty and must be reinterpreted as generalized food deception. For example, the suggestion that the pink-flowered deceptive orchid *Dactylorhiza lapponica* is a mimic of a sympatric *Pedicularis* species (Neiland and Wilcock 2000) cannot be substantiated if the entire *Dactylorhiza* lineage has pink flowers. The pink flower color in this example may have some functional significance for attracting pollinators that visit *Pedicularis*, but does not in itself indicate mimicry. Some early studies interpreted the circumboreal orchid *Calypso bulbosa* as a mimic of various rewarding species, but it is now believed to be a GFD species (Boyden 1982). Color similarity between a GFD species and a rewarding species may certainly enhance visitation in some parts of the range of the GFD species (Johnson et al. 2003b), but this "facultative" benefit (Dafni and Ivri 1981) should not be confused with the adaptive resemblance that characterizes true mimicry.

Some authors have considered generalized food deception to be a form of floral mimicry and have described it as generalized or non-model mimicry (Dafni 1984). This definition follows the logic that GFD plants "mimic" the "general search image" that many pollinators may have for large, colorful inflorescences (Menzel 1985). However, such a search image, at least in generalized pollinators, is usually highly plastic and can quickly be modified by learning (Gumbert 2000). This is particularly the case for the pollinators of GFD plants, which are often generalist foragers. Pollinators in mimicry systems, by contrast, are usually relatively specialized, although some do have the ability to develop fine-tuned preferences for particular signals through learning (Dötterl et al. 2011; Milet-Pinheiro et al. 2012; Newman et al. 2012). Despite these differences between generalized food deception and mimicry, there is increasing realization that generalized food deception (characterized by the lack of any detectable model and generalized pollination) and Batesian food-source mimicry (one model species and specialized pollination) may occupy the opposite ends of a spectrum. In between are systems of guild mimicry involving species that imitate entire guilds of co-flowering species (Dafni 1983; Bernhardt and Burns 1986; Dafni and Calder 1987; Indsto et al. 2006). Floral guild mimicry is arguably most similar to defensive "imperfect mimicry" arising from the imitation of multiple models (Sherratt 2002; Gilbert 2005).

Despite some exceptions, most floral mimics are rewardless. Indeed, floral mimics often evolve from rewardless plants. Rewardlessness is usually associated with low attractiveness of a flower and comparatively low pollination success (Willson and Ågren 1989; Jersáková et al. 2006a; Scopece et al. 2010). Therefore, rewardless flowers may be selected to evolve mimicry if circumstances allow, thus increasing the likelihood of pollinator visitation. Indeed, studies of orchid lineages in Europe, Australia, and South Africa show that both food and sexual mimicry evolved from generalized food deception (Inda et al. 2012; Johnson et al. 2013; Weston et al. 2014). Several studies have shown that the likelihood of a rewardless plant being visited

by a pollinator depends on its similarity in color to a rewarding plant previously visited by that pollinator (Fig. 3.1) (Gumbert and Kunze 2001; Gigord et al. 2002; Johnson et al. 2003b). These studies reveal a pathway whereby rewardless flowers may be selected for mimetic resemblance to co-flowering potential model species. Such selection may not necessarily lead to the evolution of mimicry (and indeed seldom does) because co-flowering communities can be variable in space and time, thus balancing out net selection pressure. For mimicry to evolve from generalized food deception, communities must be highly stable and predictable over time—which is seldom the case in temperate environments.

The role of learning

Learning is a central aspect in the evolution of all mimicry and GFD systems (Wickler 1968). In protective mimicry operators learn to avoid the model, and generalize this avoidance to the mimic, which thus gains protection. In inviting mimicry, such as floral mimicry, the mimic produces signals similar to a model to attract the operator, and avoidance learning by the operator needs to be delayed for as long as possible. Similarly, in generalized food deception, avoidance learning by the pollinator leads to a decrease in pollination success, since pollinators quickly abandon non-rewarding flowers (Dafni 1984; Internicola and Harder 2012). The patterns of learning by pollinators thus have important consequences for the maintenance of model–mimicry associations. In addition, learning is thought to influence variability in floral traits within populations of GFD plants through negative frequency-dependent selection (Smithson and Macnair 1996; Roy and Widmer 1999). Because honey bees and bumblebees have long been used as model organisms for learning in insects (Lunau et al. 1996; Chittka et al. 1997; Gumbert 2000), extensive information directly pertinent to understanding the role of innate versus learned behavior in the evolution and ecology of GFD species is now available.

Many flower-visiting animals are excellent learners, quickly associating floral signals with the presence or absence of rewards (Dyer and Murphy 2009). Learning assays utilizing the proboscis extension

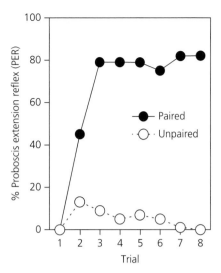

Figure 3.2 Demonstration of associative conditioning in honey bees by using the proboscis extension reflex (PER). Bees of the "paired" group received a conditioned (odor) and unconditioned stimulus (sugar solution) in close temporal association, whereas the "unpaired" group received temporally dissociated stimuli. This experiment shows that close temporal association of stimuli, but not the presentation of the stimuli alone, leads to successful conditioning. Adapted from Giurfa and Sandoz (2012).

reflex (Fig. 3.2) or with free-flying insects (Giurfa and Sandoz 2012) have been used to assess the learning capabilities of flower-visiting insects. In a classic conditioning assay pioneered by von Frisch (1919), a conditioned stimulus (scent, color, pattern) is presented together with an unconditioned stimulus (e.g., sugar water). Such conditioning assays allow analysis of the speed of associative learning, the ability of an insect to discriminate between rewarded and unrewarded stimuli, and its capacity to generalize among different but similar stimuli. Associative learning is often very rapid, for both olfactory and visual signals (Fig. 3.2), with a single conditioning trial sometimes being sufficient for an insect to learn the association between a particular stimulus and food (Menzel et al. 1993; Weiss 2001; de Ibarra et al. 2014). A pollinator may retain this information throughout its life (Giurfa and Sandoz 2012). Most flower-visiting insects have exceptional abilities to discriminate between different stimuli (Box 3.1). In terms of olfactory discrimination, Vareschi (1971) showed that honey bees can discriminate between more than 95% of the

approximately 1800 scent compounds tested. Honey bees have been shown to be able to discriminate not only between flowers of different plant species but even between different plant cultivars by using their distinct odor blends (Wright et al. 2002). Apart from needing to discriminate, pollinators also need to generalize, that is, to decide—when (variable) signals are encountered one after the other—which signals are alike (Wright and Smith 2004). Thus, this form of cognitive generalization is a basis for floral constancy, the repeated visitation of flowers of the same type while ignoring other rewarding flowers nearby. Karl von Frisch (1919) had already observed that honey bees, when conditioned to a particular odor, also responded to others that he judged as having a similar smell. Later it was shown that honey bees do indeed perceive odor molecules similarly when these molecules share functional groups or carbon-chain lengths (Fig. 3.3), and that such perceptual similarity is a proximate basis for generalization (Sachse et al. 1999; Giurfa and Sandoz 2012).

Besides odor, bees are able to generalize between similar colors. After training to a particular color, bumblebees prefer the color most similar to the trained color, but if colors are too different bees do not generalize but preferentially choose the innately preferred color (Gumbert 2000). Many earlier experiments investigating color choices used absolute conditioning (training with a single rewarded stimulus). A more biologically realistic approach is to use differential conditioning, in which an animal learns to associate one color with a reward in the presence of another color that is unrewarding. Differential conditioning experiments can show surprising results. Honey bees, for example, will sometimes choose a color that they had learned to avoid during differential conditioning over a new color (Dyer and Murphy 2009). This helps to explain why insects continue to visit GFD species in plant communities.

To understand how generalized food deception functions, it is critical to understand the rules that dictate when bees transition from one species to another. Transition can be predicted by the time bees spend on individual flowers, which probably reflects the availability of rewards, but also other factors that influence energetics, such as handling time and flight distances between flowers. The likelihood that bumblebees will transition

Box 3.1 Color vision and color discrimination

A key issue in mimicry is the ability of an operator to discriminate between model and mimic, because this will ultimately result in selection on traits that confer resemblance of the mimic to the model. The response of the operator is determined not only by its sensory system, but also by speed–accuracy trade-offs (Chittka et al. 2009). For example, a pollinator may forgo discriminating against non-rewarding mimics in order to maximize the speed at which it is able to find rewarding flowers.

Operators can potentially use a number of sensory modalities to discriminate between models and mimics. Human observers tend to emphasize visual cues, and these have been relatively well investigated. Floral mimics often match the floral color of their models strikingly well in terms of human perception. Caution is necessary, though, because the visual system of pollinators differs from that of humans. Flowers that, to humans, look different from each other may in fact be difficult for pollinators to discriminate and, conversely, flowers that look the same may in fact be easily discriminated by pollinators. For example, the deceptive orchid *Cephalanthera rubra* has purple flowers, and shares its pollinator with the blue flowers of the genus *Campanula*. Nilsson (1983b) analyzed the spectral composition of model and mimic in this system and concluded that the color match is in fact much closer when seen through the eyes of pollinators because bees typically have low receptor sensitivity for the red component of the reflectance spectrum.

Color vision, however, is determined not only by the spectral sensitivity of color receptors but also by post-receptor mechanisms that code spectral information, as well as a higher neuronal perceptual mechanism in the brain (Kelber et al. 2003; de Ibarra et al. 2014; Renoult et al. 2015). In addition, achromatic signals, such as green contrast, can be used to discriminate objects which have non-detectable chromatic properties (Giurfa et al. 1997). The best-studied insect in terms of color vision is the honey bee, which has three different color receptors, with maximum sensitivity for short- ("S", UV), medium- ("M", blue), and long- ("L", green) wavelength light. Color receptors are responsible for absorption of light quanta in specific wavelength ranges and their output is thought to be processed through opponent coding (subtraction), making chromatic contrasts detectable. For color discrimination, the output of opponent coding neurons is compared in the brain.

A range of color vision models have been developed that allow one to estimate the ability of an animal to discriminate

between two color stimuli, given an input of measured spectral reflectance values. These models plot color reflectance in a "perceptual color space," and the distances within this space can be used to predict how well color stimuli can be discriminated (Fig. 1) (Spaethe et al. 2001). The different models make different assumptions about the mechanisms of color coding and the limits of color discrimination (Telles and Rodriguez-Girones 2015). The three most widely used models are the color opponent coding model (COC; Backhaus et al. 1987) developed for the honey bee, the receptor noise-limited opponent coding model (RNL; Vorobyev and Osorio 1998) developed for a wide range of visual systems, and the color hexagon (CH; Chittka 1992) developed for trichromatic hymenopterans.

In addition to color, a pollinator's perception of a whole flower can be imitated by multispectral imaging (Vorobyev et al. 1997), giving a rough estimate of how a flower will be perceived by an insect. To do this, intact flowers are photographed with a UV-sensitive camera through five chromatic filters to allow accurate reconstruction of spectral reflectance. Reflectance spectra are then used to calculate signals of S-, M-, L- receptors in each pixel, which are subsequently converted into red–green–blue (RGB) values to display the

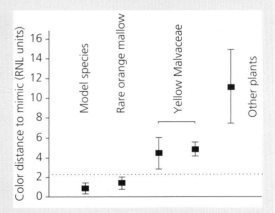

Figure 1 Mean color distances, calculated with the RNL model, between a floral mimic (*Turnera sidoides* with light orange flowers), its proposed model, (*Spheralcea cordobensis*), another rare orange-flowered mallow (*Abutilon pauciflorum*), two yellow colored mallows (*Modiolastrum malvifolium, Sida argentina*), and other co-occurring flowers. The dotted line indicates the threshold value of 2.3 RNL units below which color discrimination is suggested to be impossible for bees. Adapted from Benitez-Vieyra (2007).

Box 3.1 *Continued*

(a)   (b)

Figure 2 Floral displays of the mimic *Turnera sidoides* (B) and its model *Spheralcea cordobensis* (A) seen through a human eye (first row) and through a bee's eye (second and third rows) using multispectral imaging. The third row shows the low spatial resolution of a bee's eye at a distance of 6–9 cm from the flower. Picture © Santiago Benitez Vieyra.

flower in "bee colors"; to visualize the optical resolution of an insect's compound eye, which consists of individual optical units called ommatidia, images are projected onto ommatidial lattices to calculate ommatidial quantum catches (Fig. 2).

Despite the versatility of these models (Maia et al. 2013), it should not be forgotten that their performance may vary, thus limiting their usefulness. An empirical examination of the predictive value of the three models of color discrimination by bumblebees showed that none of the three models predicted color discrimination precisely. Despite being widely used, the CH model actually performed worst (Telles and Rodriguez-Girones 2015). Poor model performance can be due to a number of factors, including bees using both chromatic and achromatic channels for spectral discrimination, lack of inclusion of additional parameters for the specific visual system of a pollinator, and even the way that spectral reflectance from flowers is measured (de Ibarra et al. 2014). The best way to determine whether pollinators can discriminate between two differently colored objects is behavioral assay with the pollinator of interest, either with free-flying pollinators, using the interview technique, or insects that are conditioned and tested in a cage, using either natural or artificial stimuli (see Box 5.2).

---

to another species increases with the number of consecutive flower probes that are shorter than the lower quartile of flower handling time (Fig. 3.4) (Chittka et al. 1997). If bumblebees transition from a rewarding species they are likely to visit another with a similar color, but if none is available they revert to innate preferences (Gumbert 2000). Thus GFD species may benefit from both learned and innate behavior even when bees are experienced. Queen bumblebees often carry large numbers of pollinaria of GFD species (Nilsson 1983a; Fritz 1990; Cozzolino et al. 2005), indicating that they can be deceived repeatedly. Some authors have interpreted single pollinarium loads on bees as evidence that they quickly learn to avoid GFD species (Boyden 1982), but this should be interpreted with caution as there can be mechanical reasons why

insects do not accumulate multiple pollinaria on their bodies.

Innate preferences

Innate preferences are probably common among pollinators (Lunau and Maier 1995). Even highly generalist foragers such as honey bees, which readily learn to use new food sources, may have "innate search images", and not all stimuli are learned at the same speed (Menzel 1985; Smith 1991; Giurfa et al. 1995). Innate preferences by bumblebees apply not only to the color of the flowers as a whole but also to antennal responses toward contrasting "nectar guides" (Lunau et al. 1996). During natural foraging, however, innate preferences may have limited importance for flower choice because

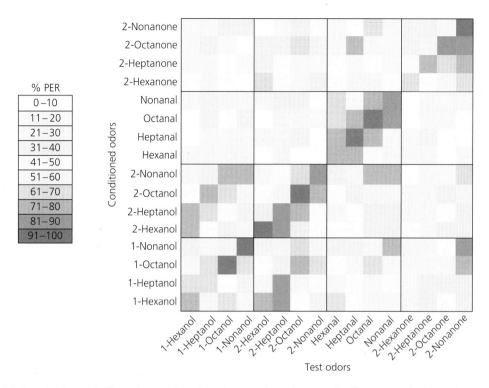

Figure 3.3 Generalization matrix of honey bees conditioned to odors as conditioned stimuli (CS). Different gray-scales indicate the percentage of bees responding to a test odor with the proboscis extension reflex. Only those bees that responded to the CS at the third conditioning trial were used ($n = 1457$). Left: percentages recorded. Right: grey-scale coded graphic display grouping the level of responses in ten 10% response categories. The data show that bees generalize odors to some degree, i.e., they respond to odors similar to those on which they have been conditioned. Adapted from Guerrieri et al. (2005).

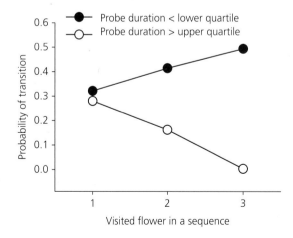

Figure 3.4 Rules for bumblebee transitions between plant species. Transitions to other species are predicted by the number of consecutive flower probes with a duration of less than the lower quartile and the number of consecutive flower probes with a duration greater than the upper quartile. The duration of flower probing is assumed to reflect nectar availability. The example given is for *Lotus corniculatus*. After Chittka et al. (1997).

learning may quickly override them (Kelber 1996; Goyret et al. 2008; Riffell et al. 2013). However, bees tend to revert to innate preferences when choosing among new flowers that are very different in color from those on which they were conditioned (Gumbert 2000). This means that plants with generalized food deception can alternatively exploit the innate preferences and conditioned preferences of experienced pollinators. Thus the ideal flower color for generalized food deception would be one that is both innately attractive to pollinators and similar to that of rewarding plants in the same community (Box 3.1). Studies of bumblebees have identified the spectral purity of the flower as an important predictor of initial long-distance attraction to naïve individuals (Lunau 1990). Similar rules may apply to olfaction. For example, the hawk moth *Manduca sexta* has innate preferences for the scent bouquets of moth-adapted flowers, and continues to visit these flowers even when conditioned to feed on bat-pollinated *Agave* flowers with a different scent profile (Riffell et al. 2013). These examples illustrate that conditioning seldom extinguishes the innate preferences of pollinators.

Avoidance learning

As pollinators learn to visit rewarding flower types they also learn to avoid non-rewarding flowers by recognizing the signals associated with rewardlessness (Fig. 3.5) (Kelber 1996; Smithson and Macnair 1997; Simonds and Plowright 2004). Negative associative conditioning of pollinators is therefore of key importance for the evolution of rewardless flowers as well as for floral mimicry, because it determines the ability of pollinators to associate particular (and often variable) plant traits with rewardlessness. An excellent learner may not be a good pollinator for rewardless plants, as it will quickly abandon them after a few trial visits. Indeed, honey bees, with their excellent learning capabilities, have rarely been documented as important pollinators of rewardless plants. Learning abilities differ among groups of pollinators. Some solitary bees, for example, are less efficient learners than are social bees (Fig. 3.6) (Dukas and Real 1991). The diurnal sphingid moth *Macroglossum stellatarum* was also shown to require longer times for associative learning than honey bees (Kelber 1996), while other insects are entirely unable

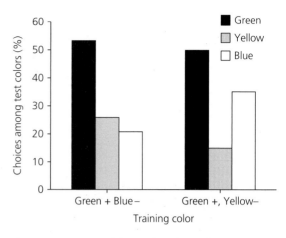

Figure 3.5 Percentage of choices of unrewarding flower colors by the moth *Macroglossum stellatarum*. The moths were previously trained to the rewarding color green, and had to choose among green, unrewarding colors (blue or yellow), and a novel color (yellow or blue). Green is always preferred, and unrewarding colors are visited less frequently. Adapted from Kelber (1996).

to learn color stimuli that differ from their innately preferred color (Lunau 1992).

Patterns of learning ability and speed are thought to have important consequences for the maintenance of generalized food deception as well as floral mimicry. A pollinator with poor discrimination abilities may be an effective pollinator of GFD plants because it may visit several non-rewarding

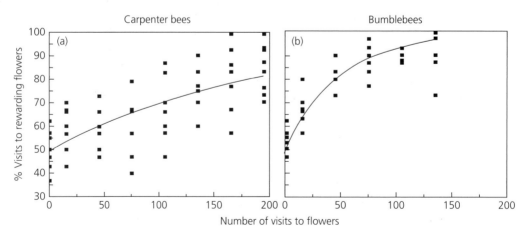

Figure 3.6 Proportions of visits to rewarding flowers by carpenter bees and bumblebees when both rewarding and non-rewarding flowers are present. Rewarding flowers are white; non-rewarding flowers are painted yellow. Bumblebees learn to selectively visit rewarding flowers more quickly than do carpenter bees. Adapted from Dukas and Real (1991).

individuals before avoiding a particular type of flower. In floral mimicry systems, poor learning abilities of pollinators will allow for loose trait matching between models and mimic. Learning abilities of pollinators will also affect the frequency of mimics in relation to models, with poor learning likely to allow for higher frequencies of mimics. Learning may also underlie patterns of variability within populations of GFD plants (see "Flower polymorphism and negative frequency-dependent selection").

Deception of nectar- versus pollen-seeking animals

Animals seeking different food types would be expected to use different floral signals. For example, nectar is seldom visible and insects seeking this resource must use cues that predict its presence, while pollen is sometimes conspicuous with a spectral pattern that contrasts with the perianth, and can thus act as a direct cue for animals. GFD species that attract nectar-seeking animals typically have a floral spur. The spur is unlikely to be a long-distance visual cue, but it is probed by insects in the expectation of finding nectar. Bees will even chew through the side of the spurs of some deceptive orchids if they perceive that they may contain nectar that is out of reach (Neiland and Wilcock 2000).

Many insects seek pollen from flowers, thus plants can use pollen imitation to signal the presence of pollen dishonestly (Plate 1). Pollen imitation is probably widespread (Osche 1979; Lunau 2000) and includes cases of intersexual mimicry where certain structures of female flowers are modified to resemble the pollen-bearing apparatus of male flowers. These cases are often classified as intraspecific or auto-mimicry, as model and mimic belong to the same species (Wickler 1965). However, in most cases pollen mimicry is relatively unspecific and can be considered as part of a strategy of GFD or generalized mimicry (Heuschen et al. 2005). There is evidence that pollinators such as bumblebees and hoverflies have an innate preference for pollen-like signals (Lunau 2000). In contrast to nectar, pollen is often directly visible to pollinators and the visual signals of pollen are well conserved. Pollen is typically yellow and absorbs UV light, due to flavonoid pigments that protect DNA against UV radiation

(Osche 1979; Lunau 2000). Pollen may also have a characteristic scent, but we know little about how pollinators use olfactory signals when searching for pollen (Dobson and Bergstrom 2000; Knauer and Schiestl 2015). Flower visitors seeking pollen may often have a characteristic search image, either through innate or acquired preferences. Bumblebees have innate preferences for pollen signals such as yellow with UV absorption, or yellow spots on a contrasting (blue) background (Lunau and Maier 1995; Heuschen et al. 2005).

Pollen is sometimes concealed to protect it from rain, UV radiation, or excessive collection by insects, and is therefore not visible to foragers. Other flowers (such as those of orchids) typically have pollen packed in inaccessible pollinia (or "pollinaria" if the pollen mass contains other accessory parts) as a reward. For such flowers, pollen imitation may provide an opportunity to increase their attractiveness to pollen-seeking foragers. Some flowers have evolved secondary pollen signals, such as enlarged UV absorption in the center of a flower or on the labellum (Fig. 3.7) (Peter and Johnson 2013).

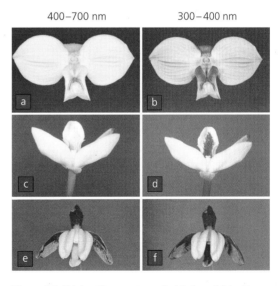

Figure 3.7 UV-absorbing patterns on the labellum of *Eulophia* orchids are believed to attract pollen-seeking solitary bees through their general spectral similarity to pollen. Images on the left are taken using human-visible wavelengths; those on the right using UV wavelengths: (a), (b) *Eulophia speciosa*; (c), (d) *Eulophia ovalis*; (e), (f) *Eulophia parviflora*. Adapted from Peter and Johnson (2013). Photos © Craig Peter.

A comparative study has shown that within flowers or flowerheads, the color of the central part is usually less diverse than that of the outer part, with the central part often similar to the color of pollen (UV-absorbing yellow) (Heuschen et al. 2005). A central, pollen-imitating spot on artificial flowers was also shown to be attractive to bumblebees; the size of the spot and its similarity to the color of real pollen was correlated with the attractiveness of the artificial flower (Heuschen et al. 2005). In addition, a yellow spot on the lower side of a corolla decreases avoidance learning of unrewarding artificial flowers by bumblebees (Fig. 3.8) (Pohl et al. 2008). Pollen-mimicking spots thus seem to increase the attractiveness of flowers and to decrease

the efficiency of bumblebees in exploiting them, to the advantage of the plant.

Another form of pollen imitation, which is more like generalized deception than mimicry, is associated with heteranthery, where showy yellow, pollen-containing "feeding anthers" attract pollinators, whereas "fertilizing anthers" are inconspicuous or have the same color as the perianth (Vogel 1978). Fertilizing anthers are often situated where they are likely to contact the pollinator on parts of its body that are inaccessible to grooming.

Pollen imitations can contain some pollen or other rewards, but can also be fully rewardless, for example when "feeding" anthers contain no pollen (Vogel 1978). In orchids, rewardless pollen mimicry is common, for example through yellow hairs or spots on the labellum, yellow staminodes, or the production of "pseudopollen" (Vogel 1978; Pansarin 2008; Shi et al. 2009; Davies et al. 2013; Duffy and Johnson 2015). Pseudopollen may in fact be rewarding, such as in *Maxillaria* spp. in which the pseudopollen of some species contains nutritious protein bodies and starch grains, whereas other species are rewardless and produce non-nutritious pseudopollen (Davies et al. 2013). Pollen mimicry is further evident in dioecious plants, in which females are usually rewardless for pollen-collecting visitors, unless they bear vestigial androecia with non-fertile but nutritious pollen (Lunau 2000) (see Chapter 7). In some *Begonia* and Cucurbitaceae species, female flowers have evolved anther-mimicking stigmas (Ågren and Schemske 1991).

Selection for showy floral displays

If generalized food deception relies on exploratory visits by insects that innately associate large flowers with a nectar reward, then it seems likely that this pollination system is associated with the evolution of unusually showy inflorescences. Since pollinators probe fewer flowers on non-rewarding inflorescences, the evolution of large floral displays in GFD species may be less constrained by the costs of between-flower selfing that are faced by rewarding species (Johnson and Nilsson 1999). Furthermore, while Batesian food-source mimics are

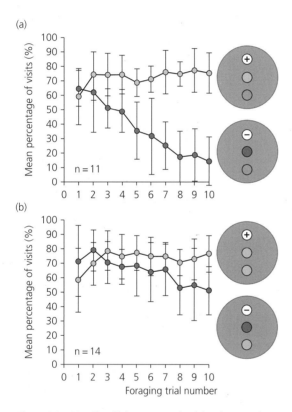

Figure 3.8 Visits of bumblebees to rewarding (+) and unrewarding (−) artificial flowers presented together in a flight cage. (a) Avoidance learning is rapid when the unrewarding flowers lack a yellow spot. (b) Avoidance learning is slower when the unrewarding flowers exhibit a yellow, pollen-mimicking, spot on a blue corolla. Color codes: light grey: yellow, medium grey: blue, dark grey: orange. Adapted from Pohl et al. (2008).

restrained by a general requirement to match the inflorescence dimensions of their models (Johnson et al. 2003a), GFD species should not have this constraint. It could even be that the appeal of orchids to humans is largely because this plant family is dominated by GFD species with large showy inflorescences. There are many studies showing that selection favors larger number of flowers in GFD orchids (Sletvold et al. 2010; Sletvold et al. 2013). However, flower number is also a target of selection in rewarding orchids (Maad 2000) and no study has specifically compared the strength of selection on display size across representative samples of deceptive versus rewarding orchids. Selection on total flower number also needs to be viewed in light of the potential future costs of such investment. Such long-term costs have been reported for several orchids (Montalvo and Ackerman 1987), but in a study of *Orchis purpurea* there was no detectable effect of total flowering display on the likelihood of flowering in a subsequent year (Jacquemyn and Brys 2010). However, the likelihood that any given flower will set fruit (proportional fruit set) is often unrelated to the total floral display, suggesting that production of extra flowers does not usually yield greater female fecundity per flower invested.

Positive directional selection on plant height has been reported in many plant studies, and the idea that this trait and generally those that promote conspicuousness may be under especially strong selection in pollen-limited GFD species has received empirical backing (Sletvold et al. 2010). In *Dactylorhiza lapponica*, however, selection favors taller plants only when there is tall surrounding vegetation (Sletvold et al. 2013).

Flower polymorphism and negative frequency-dependent selection

Intraspecific variation is common in plants, and is often especially pronounced among floral traits (Kay 1978). Some rewardless plants show spectacular floral variation, with distinct color morphs (Petterson and Nilsson 1993; Koivisto et al. 2002; Tremblay and Ackerman 2007) or fragrance types (Ackerman et al. 1997; Salzmann

et al. 2007a) present within populations. A recent study comparing floral variation in rewardless and rewarding plants (without distinguishing between deception and mimicry) found that flowers pollinated by deception are more variable than rewarding ones in morphology and fragrance (Ackerman et al. 2011). In principle, two forms of polymorphism can be found: (1) discrete polymorphism, with two or more clearly distinctive morphs, and (2) continuous polymorphism, without any distinguishable morphs but with high inter-individual variability. Polymorphism can be caused by several different evolutionary mechanisms; for example, high variability in the sense of being dissimilar from conspecifics may be advantageous for non-rewarding plants, as it may delay generalization and avoidance learning by pollinators. Alternatively, variability may be the consequence of a non-adaptive evolutionary process, such as genetic drift. Genetic drift may indeed be important for population differentiation in some deceptive plants (Tremblay and Ackerman 2001); however, overall population differentiation tends to be lower in deceptive than in rewarding orchids (Cozzolino and Widmer 2005), which does not support the idea of a prominent role for drift in shaping among-population variation in non-rewarding plants. Here we focus on pollinator-mediated mechanisms that may select for increased variability in floral traits, a topic that has received considerable attention during past decades.

As discussed earlier, pollinators usually show some form of avoidance learning after visiting rewardless flowers, by associating floral signals with rewardlessness (Kunze and Gumbert 2001). This form of generalization is thought to be delayed by high variation in floral signals (Heinrich 1975); thus, high floral variation may be an advantage for rewardless plants. This hypothesis suggests that deceptive plants should benefit from high intrapopulation variability in traits that could be used by pollinators to generalize non-rewarding floral types and subsequently avoid them. Although intuitively logical, this hypothesis has not gained much empirical support from manipulative studies on floral color (Smithson et al. 2007) and scent (Salzmann et al. 2007b; Juillet et al. 2011) that experimentally increase or decrease variability in a given trait thought to be important for generalization by pollinators. Thus there is currently no strong evidence that high variability among individuals increases reproductive success in orchids (Juillet and Scopece 2010).

A special phenomenon of pollinator-mediated selection for/against polymorphism is frequency-dependent selection (FDS). In FDS, the fitness of a morph is a function of its frequency (its commonness in relation to other morphs) in a population. FDS is therefore only likely to act on discrete polymorphisms where one morph can be distinguished by the pollinator from at least one other morph within the population. In positive FDS, the more common the morph, the higher its fitness, whereas in negative FDS, the rarer a morph is in a population the higher its fitness (Smithson and Macnair 1996). Pollinators that visit rewarding flowers often establish floral constancy, a short-term specialization on a particular flower type, leading to positive FDS because the signals of the commoner morph are more likely to be learned and thus subsequently visited. On the other hand, rewardlessness is thought to lead to negative FDS because a common morph will be more quickly avoided, and a rare morph will be more commonly visited than predicted from its frequency (Smithson and Macnair 1996, 1997). Negative FDS is expected to lead to the maintenance of stable, discrete polymorphisms in natural populations of GFD plants. By contrast, in floral mimicry, negative FDS will affect the frequency of mimics in relation to models, which usually involves two different species rather different morphs within one species (see Chapter 1 for a discussion on frequency dependence in mimicry).

In generalized food deception the potential for negative FDS, as well as its pattern in natural populations, has been examined extensively, with mixed results (Smithson 2001). In theory, the persistence of two or more color morphs is a smoking gun for a process of selection, since the process of genetic drift should always lead to fixation of one morph in the absence of selection, or even if there is weak selection. There is good evidence that bumblebees forage on arrays of artificial flowers in a way that would predict negative FDS on color morphs in natural populations (Smithson and Macnair 1997). Given the potential for negative associative conditioning, it would be expected that the rarer morph in a natural population of a species with generalized food deception should have higher pollination success than the more common morph. This can be relatively easily investigated in orchids, where pollinator limitation is often severe and thus fruit set corresponds well to pollination success. Indeed, one study in the field found evidence for negative FDS in *Dactylorhiza sambucina*, a rewardless orchid with yellow and purple morphs (Gigord et al. 2001). In this study, natural pollination was examined in artificial plots of translocated plants, with varying frequencies of yellow and purple morphs. A negative correlation between morph frequency and male and female reproductive success (RS) was found. Using a regression of fitness (i.e., RS) against color morph frequency, the study predicted a fitness equilibrium stage with a frequency of yellow morphs of between 60% and 70%, which was surprisingly well matched in 20 populations in France having an average yellow morph frequency of 69%.

The majority of subsequent studies on *Dactylorhiza* were, however, unable to confirm that patterns of fruit set are predicted by negative FDS in natural populations of GFD plants with color morph polymorphism (Fig. 3.9) (Koivisto et al. 2002; Pellegrino et al. 2005a,b; Jersáková et al. 2006b; Dormont et al. 2010). Similarly, in tropical orchids, no association has been detected between morph (both color and scent) frequency and pollination success (Ackerman et al. 1997; Aragón and Ackerman 2004; Ackerman and Carromero 2005; Tremblay and Ackerman 2007).

Taken as a whole, the evidence for an adaptive significance of flower polymorphism in GFD species is not convincing. There is neither good support for negative FDS acting on discrete polymorphism in natural populations nor evidence that deception could promote the evolution of overall variability. Nevertheless, comparative approaches suggest higher continuous variability in deceptive plants (Salzmann et al. 2007b; Ackerman et al. 2011). Therefore, non-adaptive evolution seems to be a more plausible explanation for such floral variability in deceptive plants. For example, the lack of a floral reward may relax stabilizing selection on floral traits

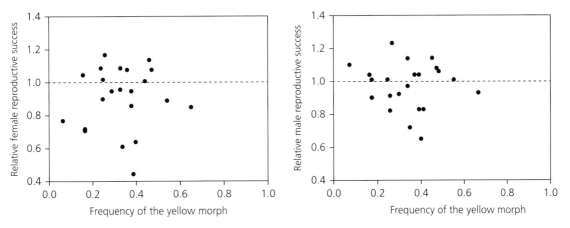

Figure 3.9 Relative female and male reproductive success of the yellow morph of the food-deceptive *Dactylorhiza sambucina* in relation to its frequency (relative to the purple morph) in 22 different populations in the Czech Republic. Negative frequency-dependent selection would suggest that yellow morphs are more successful when rare in a population, but such an association was not found in the examined populations. Adapted from Jersáková et al. (2006b).

because pollinators do not establish floral constancy but quickly avoid unrewarding plants (Juillet and Scopece 2010). Floral constancy in rewarding plants is expected to impose stabilizing selection on floral traits, because a plant that is too different from the search image of a pollinator will not be generalized as being the same and thus be visited less, and rare variants may disappear from a population.

The early flowering hypothesis

Temperate orchids that deploy generalized food deception tend to flower earlier than those that have rewards. This holds true even when analyses control for altitude and phylogenetic affinity (Fig. 3.10) (Pellissier et al. 2010). Many GFD orchids pollinated by queen bumblebees emerge early in spring at around the same time as the queens are emerging from overwintering and seeking new food sources (Nilsson 1980, 1983a). However, there are other GFD orchids that flower later in the summer and also successfully attract pollinators (Juillet et al. 2007).

Experiments have shown that if deceptive plants are dissimilar in color to other plants in the same community then earlier flowering would be favorable, but if they are similar then later flowering would be advantageous (Fig. 3.11) (Internicola

et al. 2008). The basic premise of the early flowering hypothesis for food-deceptive orchids is that early flowering plants would benefit from exploratory visits by bees that have not developed an aversion to the orchids or developed foraging constancy on rewarding plants in the same communities.

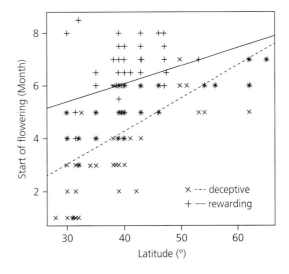

Figure 3.10 The relationship between the starting month of flowering and the southernmost latitude recorded for 233 European orchids considered in the analysis. The data show that deceptive orchids tend to flower earlier than rewarding orchids. Adapted from Pellissier et al. (2010).

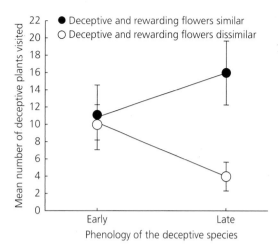

Figure 3.11 Bumblebee visits to yellow artificial inflorescences increase with time if they are conditioned on rewarding inflorescences of a similar color (dark yellow, closed circles), but decrease if they are conditioned on rewarding inflorescences of a markedly different color (blue, open circles). Adapted from Internicola et al. (2008).

In studies using arrays of artificial flowers, bees quickly learn to avoid non-rewarding morphs. However, the evidence for bees readily learning to avoid food-deceptive orchids in natural habitats is less clear. While studies show that bees are more likely to switch to other species when rewards are low or absent (Fig. 3.4) (Chittka et al. 1997), there is also evidence that pollinators can be deceived repeatedly by food-deceptive orchids. This evidence includes loads of up to several dozen pollinaria on a single queen bumblebee (Fig. 3.12) (Nilsson 1983a; Fritz 1990) and field choice experiments showing that experienced queens readily probe deceptive orchids, particularly if the orchids have a floral color similar to that of the food plant on which the bee is currently foraging (Johnson et al. 2003b).

One approach to the problem is to test whether there is selection on the date of first flowering in individual orchids. On a Baltic island, pollination success of *Orchis spitzelli* is highest before the peak in flowering of the orchid community (Fritz 1990). A similar result was obtained for *Cyripedium japonicum* in central China (Sun et al. 2009). There is also a negative effect of population size on pollination success in this species, suggesting that aversive conditioning of bees may underlie the advantage of early flowering individuals. However, Sletvold et al. (2010) were unable to detect selection acting on the start of flowering time in *Dactylorhiza lapponica*. A benefit for earlier-flowering individuals was reported for the pink lady's slipper orchid, *Cypripedium acaule*, in North America (O'Connell and Johnston 1998); however, this was attributed

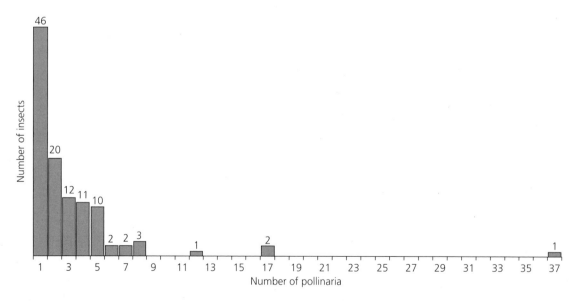

Figure 3.12 The number of pollinators carrying different numbers of pollinaria in southern Italy. Most individuals were found to carry only a single pollinarium; the maximum number of pollinaria found on a single insect was 37. Adapted from Cozzolino et al. (2005).

mainly to the effect of increased leaf cover on bumblebee abundance rather than aversive conditioning or competition from co-blooming rewarding species. The authors argued that flowering too early in spring may be selected against because of frost damage. In the food-deceptive orchid *Eulophia parviflora*, flowering of an ecotype on the coast where frost is absent occurs much earlier in spring than does flowering of an ecotype in the frost-prone inland region (Peter and Johnson 2014).

The most thorough study to date on the selective benefit of early flowering on success of GFD orchid species involved manipulation of the flowering time of *Calypso bulbosa*, a species with naturally long-lived flowers, through the use of exclusion bagging (Internicola and Harder 2012). Earlier flowering and long periods of flowering both enhanced pollination success (Fig. 3.13). Bumblebees that had experienced the orchid were less likely to probe flowers than those that were naïve, suggesting aversive conditioning, but those that had previously visited a rewarding plant species were more likely to probe the orchid (Fig. 3.14). Thus early flowering plants may benefit from naïve insects while later-flowering ones may benefit from bee conditioning

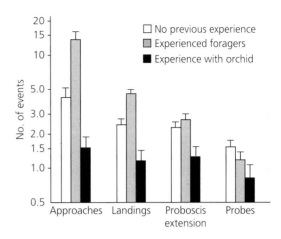

Figure 3.14 Bumblebee behavior on flowers of the deceptive orchid *Calypso bulbosa* in relation to whether they are experienced foragers, have no previous experience, or have previous experience with the deceptive orchid. Bees with experience of the deceptive orchid show the least intense responses to it. Bumblebee responses are presented on a log scale. Adapted from Internicola and Harder (2012).

on rewarding plant species (Internicola et al. 2008; Internicola and Harder 2012).

Effects of population size and plant density

More than 30 years ago Beverly Rathcke developed the idea of an optimal population size for pollination (Rathcke 1983). Rewarding plants would be expected to show a positive effect of plant aggregation (population size and density) on fecundity up to some optimum (Fig. 3.15). This component of the "Allee" effect should arise because aggregations of plants increase their attractiveness to animal foragers (a basic expectation of optimal foraging behavior) and also increase the number of mating partners (and thus conspecific pollen loads on pollinators). Rathcke argued that at some stage of increasing plant number and density pollinators should become saturated by the number of plants, leading to competition and reduced fecundity.

Facilitation and competition among conspecifics should be somewhat different for deceptive plants as there is no energetic incentive for pollinators to concentrate their foraging in such patches (Fig. 3.15). The only benefit for plants would be

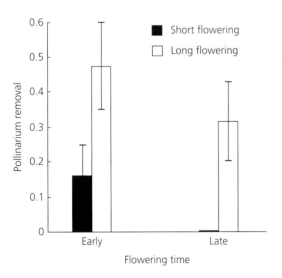

Figure 3.13 Removal of pollinaria from flowers of the deceptive orchid *Calypso bulbosa* in relation to time of flowering and flowering duration. Long flowering increases reproductive success, especially in late-flowering individuals. Adapted from Internicola and Harder (2012).

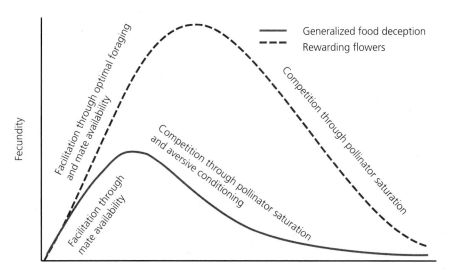

Figure 3.15 Expected relationships between population size and fecundity in deceptive and rewarding plants. Because pollinators can learn to avoid deceptive flowers, competition for pollinators is expected at lower population sizes and densities than is the case for plants with rewarding flowers.

the availability of nearby mating partners. Indeed, dense aggregations of deceptive species should increase the rate of aversive conditioning in pollinators, and the optimal level of aggregation in terms of population size and density is expected to be lower than it is for rewarding species.

Rathcke's idea of a population size-density optimum has received some support from a study of the rewarding European orchid *Listera ovata* (Brys et al. 2008), but there are still too few studies to firmly test whether optimum population sizes for food-deceptive species are lower than those of rewarding species. In a 3-year study of the circumboreal food-deceptive orchid *Calypso bulbosa*, a decline of fecundity in large populations was evident in only one of three years (Alexandersson and Ågren 1996). A general trend toward reduced fecundity in large populations was also observed for three GFD orchids on the Baltic island of Gotland, but this was only evident when data were pooled across different years (Fig. 3.16) (Fritz and Nilsson 1994). In a Belgian study, population size was positively associated with fruit set and seed quality in the rewarding orchid *Gymnadenia conopsea*, but not in the GFD species *Orchis mascula* (Meekers and Honnay 2011).

In the food-deceptive orchid *Cyripedium japonicum*, isolated plants tended to have better pollination success than those in dense patches (Sun et al. 2009).

Remote habitats versus the magnet species effect

There has been some experimental evidence for the idea that GFD plants would face competition from co-flowering rewarding plants. The "remote habitat" hypothesis (Nilsson 1992) proposes that pollination success of rewardless plants would be highest in habitats with few co-flowering species. Evidence for the remote habitat hypothesis was obtained when experimental addition of rewarding violets to rewardless *Dactylorhiza incarnata* populations reduced the pollination success of these orchids (Lammi and Kuitunen 1995). In another experiment, pollination success of the yellow morph of *D. sambucina* was negatively influenced by the aggregation of rewarding blue-colored *Muscari neglectum* (Internicola et al. 2006). This competition was likely enhanced by the dissimilar floral signals (i.e., color) of the two species.

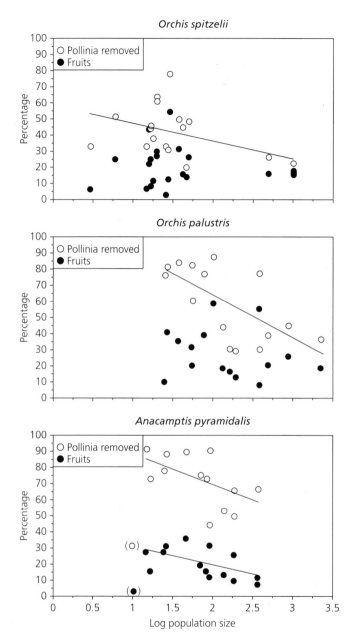

Figure 3.16 Association between population size and proportion of pollinaria removed and fruit set in different food-deceptive orchids in Sweden. In all species, population size is negatively associated with pollination success. Adapted from Fritz and Nilsson (1994).

While ecology has been dominated by a focus on competition, there is increasing appreciation of the role of facilitation among species. Rather than suffering competition, rewardless plants may in fact benefit from co-flowering rewarding plant species in terms of pollination success. For example, a benefit of growing near ericaceous shrubs was observed for the pink lady's slipper orchid, *C. acaule*, in North America (O'Connell and Johnston 1998). This kind of ecological facilitation has been called the "magnet species effect" (Thomson 1978). Magnet species create local aggregations of pollinators that benefit rewardless plants, and this works most efficiently when the magnet species resemble rewardless plants in their floral traits (Johnson et al. 2003b; Juillet et al. 2007; Pellegrino et al. 2008). This again shows potential selection for floral mimicry, but magnets differ from models in that a magnet can be any rewarding species, varying in space and time, and pollinators are not specialized toward it. Consistent selection for mimetic resemblance is only expected when pollinators are obligate specialists on a given plant species, which initially may serve as a magnet but subsequently develops into a model as the orchid develops particular adaptations for mimicry (Peter and Johnson 2008). Feinsinger (1987) argued that facilitation of one rewarding species by another would initially be positive and then become negative due to competition as the abundance of the second species increases. Available data for GFD species suggest that facilitation increases linearly with abundance of nectar plants (Fig. 3.17)

(Johnson et al. 2003b). However, facilitation was only evident when rewarding species had a color similar to that of the GFD species. Thus the benefits of magnet species can be two-fold: increasing local abundance of pollinators, and conditioning pollinator behavior.

Pollination system generalization and hybridization

An important feature of generalized food deception is an overall lack of specificity in pollinator assemblages (Nilsson 1983a; Cozzolino et al. 2005; but see Peter and Johnson 2013) and the generalist foraging behavior of the pollinators involved. The signals that are deployed by GFD species are often quite general, although the pollinator assemblages may be filtered by morphology (Bänziger et al. 2005; Li

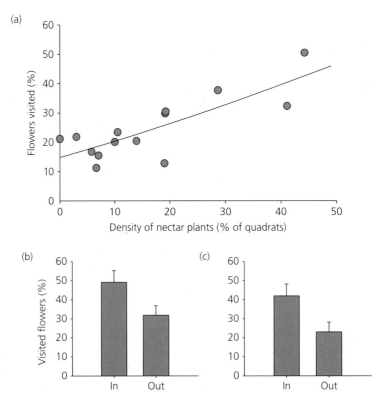

Figure 3.17 The magnet species effect in *Anacamptis morio*. (a) Association between the relative density of nectar-producing magnet plants and the mean proportion of flowers visited on *A. morio* inflorescences in 14 meadows. The lower panels show the pollination success of *A. morio* inflorescences experimentally translocated into or out of rewarding patches of *Geum rivale* (b) and *Allium schoenoprasum* (c). Adapted from Johnson et al. (2003b).

et al. 2008). Orchids that deploy this mode of deception must share pollinators with many other plants, including related orchid species. Orchids are often assumed to have a lock and key mechanism of pollen transfer, but this is actually seldom the case. In most orchids there are few mechanical barriers to pollen transfer between closely related species, and interspecific pollen flow is probably common. Pollinaria themselves are sometimes quite easy to identify when they are still attached to insects, but the actual level of pollen transfer between species is hard to quantify as the pollen masses begin to disintegrate soon after deposition on a stigma. The main evidence for sharing of pollinators is derived from direct observations of pollinator behavior (Nilsson 1983a), the presence of several types of pollinaria on insects (Cozzolino et al. 2005), and natural hybridization. In a genetic fingerprinting study conducted in Italy, Cozzolino et al. (2005) found that pollinaria of more than one orchid species were attached to one-third of 62 insects that carried two or more pollinaria (Fig. 3.18). All of these pollinaria were placed in a similar position on the head, thus revealing the potential for hybridization. Granular pollen of plants representing many families is often deposited on the stigmas of GFD orchids, but this does not usually interfere with orchid reproduction (Neiland and Wilcock 1999, 2000). Evidence for interspecific pollen transfer between orchids has also been obtained by using an electron microscope to examine the exine of orchid pollen deposited on stigmas (Neiland and Wilcock 2000), but this method is exceptionally time-consuming and not practical in the majority of cases. There is now evidence that post-pollination barriers are very important for the coexistence of food-deceptive *Orchis* species (Fig. 3.19) (Cozzolino et al. 2004; Scopece et al. 2007).

Overview and perspectives

One of the major challenges in GFD is understanding the causes of its evolutionary origin and maintenance. Because GFD plants often have very low fruit set, it is difficult to see how this pollination mode can be an evolutionary stable strategy (ESS). Several studies have addressed this problem (see Chapter 2), but a conclusive solution is still lacking. Because GFD is probably ancestral to mimicry, the transition between these pollination systems is of special interest but has as yet been little studied. Mimicry may generally increase pollination success or the reliability of pollination, but may also increase extinction risk through the typically higher pollinator specificity

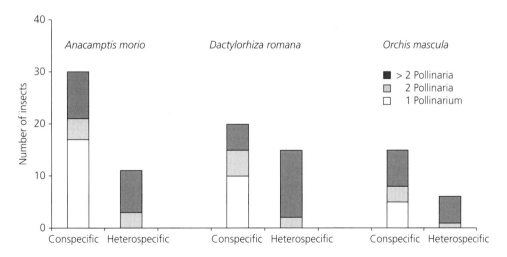

Figure 3.18 Specific and non-specific pollinators of three food-deceptive Mediterranean orchid species. The first bar shows the number of insects carrying only conspecific pollinaria. The second bar shows the number of insects carrying heterospecific pollinaria (from more than one orchid species). Overall, the specificity of insects visiting these GFD orchids is low. Adapted from Cozzolino et al. (2005).

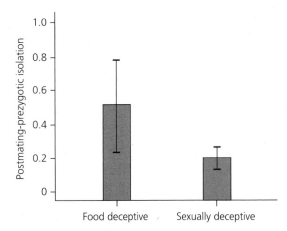

Figure 3.19 Strength of post-mating–pre-zygotic isolation (pollen–ovule interaction) in food- and sexually deceptive orchid species. Isolation was measured by crossing species with a genetic distance of 0.008–0.032. Adapted from Scopece et al. (2007).

and dependence on the occurrence of a model. The circumstances favoring the transition between GFD and mimicry, such as plant community composition and its temporal and geographic stability, are also little understood.

Overall, GFD has mostly been studied in terrestrial orchids growing in temperate regions; the majority of GFD plants likely occur in the tropics, however, and many of them are probably epiphytic. Thus to reduce this obvious bias we need more studies from tropical areas, but also more from non-orchid plant families. Also, effects of population size and/or density on the fecundity of GFD plants have hardly been taken into account, yet they may have a strong influence on selection on quantity of reward production. The same is true for the community of co-flowering plants that may act either as magnets for or competitors of GFD plants.

The common occurrence of discrete color polymorphisms in GFD plants is currently unexplained, despite the past focus on negative frequency dependence and its effect on the maintenance of polymorphism. Although several species of GFD plants show such color polymorphisms, it is as yet not clear whether such polymorphisms are indeed promoted by this pollination system and occur more frequently in GFD plants than in rewarding plants. Phylogenetically controlled analyses that include the huge diversity of tropical GFD plants are needed to shed more light on this question.

From the perspective of pollinators, learning has been investigated almost exclusively in social bees such as honey bees and bumblebees that can easily be kept in the laboratory and are thus amenable to experimental work. These are obviously not the only pollinators of GFD plants, and they may, especially in tropical areas, play a rather minor role overall. Thus, a more balanced view on learning in a greater variety of pollinator insects is highly desirable for a more comprehensive understanding of how pollinator behavior imposes selection on floral reward production and advertizing strategies (Weiss 2001).

CHAPTER 4

Batesian food-source mimicry

Introduction

Plants that lack floral rewards can attract pollinators if their flowers sufficiently resemble those of sympatric food-rewarding plants. This is because animals that associate the signals of rewarding flowers with food may also be attracted by similar signals in a deceptive species. The success of this form of Batesian mimicry depends on how easily pollinators can be duped in the first place and on their subsequent ability to learn to avoid the deceptive flowers.

The application of the concept of Batesian mimicry to plants has been controversial. Little (1983), for example, felt that the term was inappropriate for plants because the consequence of floral food-source mimicry is attraction of pollinators, rather than repulsion of predators. Others, such as Dafni (1984) and Johnson (1994), argued that the evolutionary properties of food-source mimicry in plants are similar to those of classical Batesian mimicry in animals: In both systems, the resemblance of one species to another is favored by natural selection because a signal receiver, whether predator or pollinator, perceives the mimic to be representative of the model. This cognitive misclassification of the mimic results in behavior by the operator (a signal receiver that acts on information received) that enhances the fitness of the mimic.

Most cases of floral food-source mimicry are likely to involve a strong element of operator conditioning (its "education," in the classical language of Fisher 1930). There are many studies, dating back to the classic experiments by Karl von Frisch, showing that flower-visiting insects can learn to associate particular colors, shapes, patterns, and floral odors with the presence of a reward (Barth 1991). Most flower-visiting insects have innate flower preferences that are rapidly altered by learning. The preferences of flower-visiting animals conditioned on rewarding plants can be exploited by non-rewarding mimics (Gigord et al. 2002; Newman et al. 2012). A key question is whether these exploited animals can eventually learn to avoid mimics, and whether this depends on the relative frequencies or absolute numbers of models and mimics.

In contrast to food-source mimicry, sexual and oviposition-site mimicry involve mostly innate responses to signals by operators (see Chapters 5 and 6). As a result, theoretical models for the evolution of Batesian protective mimicry (whereby mimicry confers protection to the mimic from predation), which are based on learning, are probably more easily adapted for food-source mimicry than for sexual and oviposition-site mimicry because the latter two do not assume that the initial responses of the operator toward the mimic are based on experience with the models. However, it seems likely that many specialized pollinators have at least some innate attraction to the signals of food-source mimics. This is almost certainly the case for the pollinators of flowers that mimic animals that serve as food sources for pollinators (Oelschlägel et al. 2015).

One of the major difficulties with studying food-source mimicry in flowers is to distinguish between floral syndromes and mimicry phenomena. Floral syndromes are patterns of convergent evolution that arise when unrelated plants adapt to the sensory systems, morphology, and behavior of a particular pollinator (Fenster et al. 2004). Importantly, species belonging to a pollination syndrome do not

Floral Mimicry. Steven D. Johnson & Florian P. Schiestl.
© Steven D. Johnson & Florian P. Schiestl 2016. Published 2016 by Oxford University Press.
DOI 10.1093/acprof:oso/9780198732693.001.0001

impose selection for resemblance on each other, but the selection only arises from the shared use of the same pollinators. Mimicry, on the other hand, arises when there is selection for resemblance per se. The difficulty is that floral food-source mimics have generally adapted to the same pollinators as their models and therefore belong to the same floral syndrome. The key question, therefore, is whether the similarity between plant species extends beyond that expected from a floral syndrome and is the outcome of selection for resemblance to a particular model or set of models. For example, a plant that mimics the floral food sources of butterflies would be expected not only to have the basic floral features required for butterfly pollination, such as a long floral tube (a syndrome trait), but also to imitate closely the specific signals of the model plant species (a mimicry trait).

The problem of distinguishing between syndrome and mimicry traits is particularly acute for resemblance between rewarding plant species ("Müllerian" floral mimicry); this is discussed in greater detail in Chapter 7. It is less important in the case of Batesian food-source mimicry since the mimic is deceptive and often rare, and thus likely to parasitize existing plant–animal interactions by means of mimicry.

The earliest suggestions of mimicry between flowers of different species were made during the late 1960s and early 1970s (van der Pijl and Dodson 1966; Nierenberg 1972). These consisted mostly of qualitative information about floral similarity and the sharing of pollinators between plant species

Figure 4.1 Amots Dafni of the University of Haifa, Israel, carried out pioneering studies of pollination in Mediterranean orchids and proposed a number of examples of floral mimicry.

(Boyden 1980; Beardsell et al. 1986). The first empirical evidence for floral food-source mimicry was obtained in the early 1980s (Dafni and Ivri 1981; Nilsson 1983b) (Fig. 4.1). Experimental approaches to the problem were adopted more recently (Peter and Johnson 2008; Jersáková et al. 2012; Newman et al. 2012), but there are still very few examples of Batesian food-source mimicry in plants (or indeed of Batesian protective mimicry in animals) for which all of the predictions that we outline in this chapter have been tested.

Theoretical criteria and predictions

In Chapter 1 we proposed a set of general predictions for mimicry systems. In this chapter we examine whether the available evidence supports these predictions for the case of Batesian floral mimicry. These predictions were that:

1. Mimics and models should occur in the same habitat, have overlapping phenologies, and interact with the same operators.
2. The mimic should resemble the model to the extent that operators have difficulty in discriminating between them.
3. The responses of operators to signals of the mimic should be shaped by their experience with the model.
4. The fitness of the mimic should be higher in the presence of the model than in its absence.
5. Individuals that resemble models more closely will experience higher fitness than individuals that resemble models less closely.
6. Mimics should perform best when models are relatively abundant.

And, as a consequence of all of the above:
7. The geographic distribution of the mimic should be nested within that of the model.

We also made the ancillary prediction that:
8. Mimics that extend beyond the range of a single model should exhibit geographic variation in signaling traits, either because they mimic different models in different geographical regions or because of the loss of mimicry outside the range of the original model.

This last prediction is particularly applicable to food-source mimicry. The association between Batesian mimicry and geographic polymorphism is well known in protective mimicry systems (Joron and Mallet 1998; Ruxton et al. 2004) and we return to this issue at the end of this chapter.

Throughout this book we emphasize that for a species to qualify as a mimic it must possess an "adaptive resemblance" to another species (Starrett 1993). In other words, at least some of the traits that confer resemblance between species must be novelties and have evolved for their current mimicry function. This does not exclude the possibility that a subset of the traits that contribute to the resemblance among species may have evolved prior to the ecological association and thus serve as pre-adaptations (or exaptations, *sensu* Gould and Vrba 1982). Phylogenies are now used routinely to support hypotheses about adaptation and to identify exaptations in mimicry systems (Schiestl and Dötterl 2012; Ma et al. 2016). Some floral mimicry systems have distinctive chemical signatures and can thus be reconstructed back in time (Schiestl and Dötterl 2012), but this is not usually possible for Batesian food-source mimicry due to the diversity of models (and thus floral signals) and the difficulty in inferring historical coexistence of mimic and model species. There is thus no simple means of reconstructing the history of food-source mimicry within a lineage; however, we can identify the history of individual traits that are involved in the function of food-source mimicry (Johnson et al. 2003a; Papadopulos et al. 2013; Weston et al. 2014; Ma et al. 2016).

Phenology and pollinator sharing

The most basic conditions that must be met before one species can be considered a Batesian mimic of another species are that the species should occur together, have overlapping phenologies, and interact with the same operator or operators. It has been suggested that highly mobile operators with long memories could select for mimicry in species that do not occur in the same habitat or do not occur concurrently with the models (Waldbauer 1988),

but we think that this scenario is very unlikely in the case of floral food-source mimicry. Most insect pollinators forage in small areas and have labile foraging preferences, meaning that mimics would almost certainly have to be intermingled with their models if they are to benefit from their resemblance to those models. The situation is less clear for highly mobile bird pollinators. The suggestion that nectarless forms of the North American wildflower *Lobelia cardinalis* may function as Batesian mimics of a guild of hummingbird-pollinated plants (Brown and Kodric-Brown 1979) was later criticized because this plant species is found in clumped populations, thus affording "little opportunity for alternative exposure of models and mimics to potential pollinators" (Williamson and Black 1981). This particular example is also problematic because it seems likely that the *Lobelia* species in question resembles other species simply because it belongs to the same floral syndrome of hummingbird pollination. We return to this issue later in the book (p 125) where we discuss the proposition by Grant (1966) that floral similarity among bird-pollinated flowers can be considered a form of Müllerian mimicry.

The general pattern, evident from a wide range of studies, is for Batesian food-source mimics to flower concurrently and grow intermingled with their models. There are some exceptions, such as *Cephalanthera rubra* which starts flowering some two weeks before its *Campanula* models (Nilsson 1983b). In such cases it seems doubtful that the system involves a strong learning component and it is more likely that the mimic deploys signals that elicit innate responses by pollinators.

One of the great advantages of using plants as a model system for mimicry is that it is relatively easy to observe the behavior of pollinator operators. By contrast, it is often difficult to establish whether animal models and mimics interact with the same predator operators because predation itself is seldom observed directly. Evidence for the sharing of pollinators in Batesian food-source mimicry is often obtained by direct observations and/or by capturing insects carrying pollen of the mimic while they feed on flowers of the model. Orchids, which make up the majority of Batesian food-source mimics, often have very distinct pollinaria, making it straightfor-ward to link mimic and model. Evidence for pol-linator sharing has been harder to obtain for some putative food-source mimicry systems in tropical forests. For example, the case for mimicry between orchids in the Oncidiinae and oil-producing Mal-phigiaceae seems compelling on the basis of their visual similarity, yet to date there are only a handful of reports of pollinator sharing between Oncidiinae and Malphigiaceae (Carmona-Diaz and Garcia-Franco 2009; Vale et al. 2011), despite the hundreds of plant species that are potentially involved in this system.

Advergence and function of floral signals

Flower color

Signals of Batesian mimics evolve to imitate those of their models, but the models are not under selec-tion to share signals with their mimics—if anything, selection in models will favor their signals being dissimilar to those of mimics. This pattern of unilat-eral evolution is referred to as advergent evolution (Brower and Brower 1972), to distinguish it from classic convergent evolution in which unrelated species develop similar phenotypes in response to a common environmental factor.

Human perception can be unreliable for gauging how a predator or pollinator may perceive other organisms. Similarity in signals between mimics and their models needs to be measured objectively and in terms of the sensory systems of the signal receivers. The first technique that was applied to assess visual floral signals of putative mimics and their models was UV photography using special filters that exclude wavelengths visible to humans (400–700 nm) but not the UV wavelengths (300–400 nm) that are behaviorally relevant to most in-sects. The technique was applied by Nierenberg (1972) in his study of similarity in flower colora-tion and shape between orchids in the Oncidiinae and various species of Malphigiaceae in the Car-ibbean. Nierenberg found that the central parts of the flowers of the malphigs and the orchids are both strongly UV-absorbing. This technique remains useful for revealing complex patterns of UV light reflectance among and within floral parts, but does not provide an accurate measurement of color per se.

Nilsson (1983b) pioneered the use of reflectance spectrophotometry to compare the colors of puta-tive floral mimics and their models. This method, which allows spectral reflectance of flower parts to be measured across the biologically relevant 300–700 nm range, has since been applied to a very wide range of potential mimicry systems. Data obtained from reflectance spectrophotometry can be entered into vision models to obtain a measure of similarity from the perspective of the sensory system signal receiver (see Box 3.1). A particu-larly widely used vision model is the color hexa-gon for the Hymenoptera (Chittka 1992; Chittka et al. 1992). This model appears to have wide ap-plication across a range of Hymenoptera. A "rule of thumb" useful for mimicry studies is that loci in this color space that are separated by less than 0.1 hexagon units are unlikely to be discriminated by hymenopterans (Chittka et al. 1993), but this is only a rough approximation and bioassays are usually needed to obtain certainty about whether color differences are discernible by pollinators or not. Vision models for other insects, such as flies, are less well tested in terms of their applicability across a range of taxa. Receptor noise-limited mod-els (Vorobyev and Osorio 1998) can be applied to a wide range of taxa and have been used in some mimicry studies (Benitez-Vieyra et al. 2007; Kelly and Gaskett 2014), but have drawbacks because of uncertainties around the accurate estimation of re-ceptor noise (Box 3.1).

When comparing color traits of mimics and models it is also helpful to include other species related to the mimic as well as unrelated species that occur in the same community. This allows an assessment of whether the mimic and model oc-cupy a portion of color space that is evolutionar-ily novel in the context of the mimic lineage and also whether the model and mimics occupy a color space that is distinct from the rest of the commu-nity in which pollinators make foraging choices. In the Neotropics, some yellow-flowered orchids in the Oncidiinae closely match oil-producing Malphigiaceae in terms of flower color and shape, and this has long been considered an example of mimicry (Nierenberg 1972; Vale et al. 2011). Spectral analysis has confirmed that some of these orchids and their likely models in the Malphigiaceae occupy a portion of color space (the UV–green sector in the hymenopteran color hexagon, which appears yellow to humans) that is distinct from related orchids and also from other plants in the same communities (Papadopulos et al. 2013) (Fig. 4.2). The hexagon units separating loci for the orchids and Malphigiaceae are generally smaller than the threshold value for discrimination by bees. This pattern of similarity is highly unlikely to be explained by chance. Even more compelling is that the bee UV–green color signal evolved independently in at least 14 genera in the Oncidiinae, confirming that mimicry of oil-producing Malphigiaceae had multiple origins (Papadopulos et al. 2013). Other Oncidiinae with flower colors that fall in the bee blue–green hexagon sector may use generalized food deception or be Batesian mimics of other Malphigiaceae (Carmona-Diaz and Garcia-Franco 2009). Some Oncidiinae do produce small amounts of oil (Blanco et al. 2013), so not all species that resemble Malphigiaceae are necessarily examples of Batesian food-source mimicry. However, as will be explained later in this chapter, rare plants that produce small amounts of rewards may still be considered Batesian mimics if the benefit gained from their resemblance to other species is one-sided and the evolution of signals is thus advergent. Evidence suggests that flowers of orchids in the Oncidiinae are vastly outnumbered by flowers of Malphigiaceae, and selection in the orchids, even if they do produce oil, may therefore favor mimicry of the visual displays of Malphigiaceae flowers.

The most reliable way of determining whether a trait plays a functional role in an ecological interaction is to modify the trait experimentally and observe the consequences. Matching of floral spectral reflectance of putative mimics and models, for example, suggests that color plays a role in mediating these interactions, but experimental confirmation has rarely been obtained.

The functional significance of flower color in floral food-source mimicry has been elucidated in the

South African food-deceptive orchid *Eulophia zeyheriana* that appears to mimic the nectar-producing flowers of *Wahlenbergia cuspidata* (Campanulaceae) (Fig. 4.3). The two species have closely matched spectral reflectances and share the same pollinators. Alteration of the UV reflectance of flowers of the orchid dramatically decreased its attractiveness to bees, suggesting that flower color is pivotal in this mimicry system (Peter and Johnson 2008). Subsequent experiments involving manipulation of UV reflectance in the model flowers confirmed the importance of visual cues for the attraction of bees in this system (Welsford and Johnson 2012).

There are limits to the extent to which traits of live flowers can be experimentally modified. In cases where mimics exhibit geographic polymorphism, forms can be translocated and presented in arrays at particular sites to test how traits that vary geographically influence attraction to pollinators (Newman et al. 2012). A different approach is to use artificial flowers to test the function of traits such as color and flower shape. Long-proboscid flies, for example, are easily duped by well-designed plastic flowers and probe them in a manner identical to that for real flowers. Plastic inflorescences varying in color, shape, and floral markings (Fig. 4.4) have been used to identify visual signals that mediate the attraction and probing behavior of the long-proboscid tabanid fly *Philoliche aethiopica* (Jersáková et al. 2012). This has given insights into the functional significance of floral traits of the South African deceptive orchid *Disa pulchra* (Plate 2a), which uses mimicry to exploit the relationship between *P. aethiopica* and the rewarding iris *Watsonia lepida*. In field experiments, pink and blue plastic inflorescences matching the spectral reflectance of the *Watsonia* proved highly attractive to flies, while white, yellow, orange, and red plastic flowers were rejected (Fig. 4.5). This finding suggests that matching the spectral reflectance of rewarding plants was a key requirement for this system of mimicry to evolve. Although the blue and pink plastic flowers appear quite different to humans, this is mainly due to differences in long-wavelength reflectance. In a model of fly vision, the spectra of pink and blue plastic flowers occupy the same quadrant of color space as the *Watsonia* and *Disa* and are thus probably

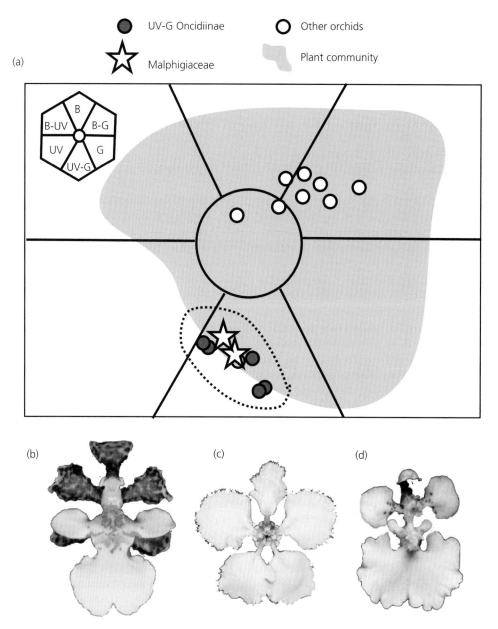

Figure 4.2 Some orchids in the Oncidiinae closely match the floral spectral reflectance of co-occurring Malphigiaceae. (a) Plants belonging to this putative mimicry system (circled with a dotted line) occupy a distinct portion of the bee color hexagon space (UV–green; see insert) relative to the rest of the flowering plant community and other related orchids. These orchids also resemble Malphigiaceae in flower shape: (b) *Trichocentrum ascendens* (Oncidiinae), (c) *Stigmatophyllon* sp. (Malphigiaceae), and (d) *Rossioglossum ampliatum* (Oncidiinae). Adapted from Papadopolus et al. (2013). Photos © Franco Pupulin.

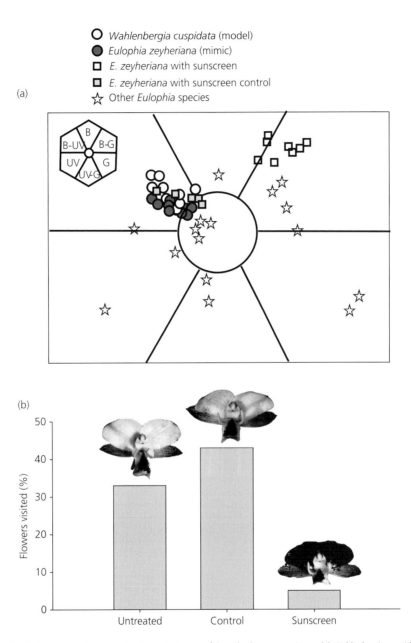

Figure 4.3 The orchid *Eulophia zeyheriana* occupies the same portion of the color hexagon as its model *Wahlenbergia cuspidata*. (a) Application of sunscreen to the orchid flowers alters their position in the color hexagon. (b) Sunscreen dramatically lowers the attractiveness of the flowers to bees compared with untreated flowers and flowers treated with a control (non-active) sunscreen solution. The difference in floral visitation between untreated and control flowers was not significant. Representative flowers shown above the bars for each treatment group were photographed in UV light. Adapted from Peter and Johnson (2008). Photos © Craig Peter.

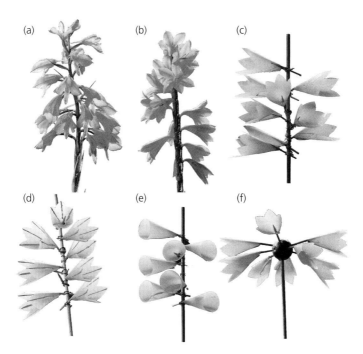

Figure 4.4 Artificial flowers can be used to identify traits that influence pollinator visitation to floral mimics. (a) *Disa pulchra*, an orchid mimic. (b) *Watsonia lepida*, the model for *Disa pulchra*. (c) Artificial (plastic) flowers. Pink and blue artificial flowers are as attractive to long-proboscid flies as are real *Disa pulchra* flowers. (d) Plastic flowers with nectar guides increase probing behavior by flies. (e) Plastic flowers lacking a dissected margin are less attractive. (f) The arrangement of plastic flowers on the inflorescence does not affect their attractiveness to flies. Adapted from Jersáková et al. (2012). Photos © Steve Johnson.

indistinguishable to flies. Nectar guides similar to those on the *Watsonia* and *Disa* were also painted onto the artificial flowers (Fig. 4.4), and although these did not affect attraction they increased the probability that the flies, once attracted, would probe the flowers.

Complex color patterns

Color signals of animals and plants often consist of complex patterns that influence the behavior of signal receivers. Mimic organisms can adopt similar patterning for their own protection or for attracting pollinators. Indeed, the remarkable similarity in patterning among animal mimics and their models, first noted by Bates and Müller, has been the principal inspiration for the past 150 years of research into the evolution of mimicry. Despite the central importance of color patterns in mimicry theory,

there are surprisingly few field-based experimental studies that have shown that similarity in pattern, not just overall coloration between mimics and models, is important for the functioning of animal mimicry systems (Brower et al. 1964; Kikuchi and Pfennig 2010). Most of the evidence for the adaptive significance of pattern in protective mimicry has been obtained from studies of trait correlations and laboratory experiments (Ruxton et al. 2004). The role of color patterns is even less well understood in plant mimicry systems, but is receiving an increasing amount of attention.

Contrasting color patterns on flowers can take the form of nectar guides on petals and sepals or differences in color between different floral organs. For example, the spectral reflectance of anthers and pollen may contrast with that of the outer perianth of a flower (Lunau 2000). There is now good evidence that pollinators use these contrasting color patterns

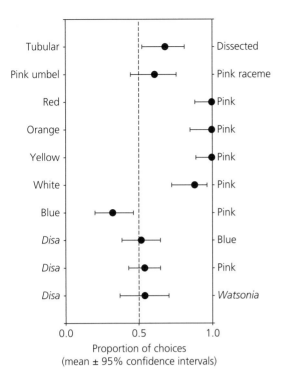

Figure 4.5 Results of experiments to test the functional significance of floral traits in a food-source mimicry system involving *Disa pulchra* (mimic) and *Watsonia lepida* (model). Colors refer to plastic raceme-shaped artificial inflorescences (see Fig. 4.4), unless otherwise stated. Experiments to test flower outline involved pink artificial inflorescences. Means with confidence intervals that do not overlap the dotted line of equal choice represent a significant preference. Adapted from Jersáková et al. (2012).

as cues for making decisions about whether to alight and probe or grasp flowers.

Similarity in color patterns between the flowers of food-deceptive orchids and co-occurring rewarding plant species have inspired a number of investigations of possible mimicry. In Australia some sun orchids (*Thelymitra* spp.) have flowers which closely resemble the purple, pink, or yellow radially symmetrical flowers of pollen-rewarding plant species, particularly petaloid monocotyledons. Not only are *Thelymitra* flowers almost perfectly radially symmetrical (a highly unusual reversal to a "retrograde" floral shape among orchids, which usually have bilaterally symmetrical flowers with a prominent labellum), but they also have central yellow or black "dummy anthers" made of a

modified column. Technically this structure is a "clinandrium" (Dressler 1981), consisting of the filament of the single fertile anther fused to the two connate staminodia. In some species the staminodia terminate in trichome brushes. The function of this floral arrangement appears to be to induce pollen-seeking insects (i.e., bees) to grasp the column, such that pollinia are deposited on the bee's underside, typically the abdomen (Bernhardt and Burns 1986). The clinandrium of some *Thelymitra* species resembles the poricidal anthers of flowers that are buzz-pollinated by bees.

Suggested models for *Thelymitra* species with blue and pink–purple flowers include various pollen-rewarding petaloid monocots (Bernhardt and Burns 1986; Edens-Meier et al. 2014), while those for yellow-flowered species are suggested to be pollen-rewarding species of *Hibbertia* (Dilleniaceae) and *Goodenia* (Goodeniaceae) (Dafni and Calder 1987). Other smaller-flowered *Thelymitra* species are known to be autogamous. Actual experimental evidence for floral mimicry in *Thelymitra* is very fragmentary, consisting mainly of morphological and color similarity and some evidence of pollinator sharing with putative models (Bernhardt and Burns 1986; Dafni and Calder 1987). Detailed spectral reflectance analyses have not been conducted for *Thelymitra* species and their putative models. The pollination systems of large-flowered *Thelymitra* species may range along the continuum between generalized food deception and Batesian mimicry. For example, generalized food deception is likely for *Thelymitra epipactoides*, which is pollinated by highly polylectic halictid bees (*Nomia* sp.) that collect pollen from many different plant species (Cropper and Calder 1990).

The food-deceptive Australian orchid genus *Diuris* contains a number of species that are strikingly similar in floral color, patterning, and shape to various legume species with unusual "egg and bacon" color patterns on their flowers (Beardsell et al. 1986). These are likely Batesian food-source mimicry systems; some evidence for this has been obtained in a study of *Diuris maculata* conducted in eastern Australia (Indsto et al. 2006). This orchid shares pollinators (small *Trichocolletes* bees) with two legume species, *Hardenbergia violacea* and *Daviesia ulicifolia*. The spectral reflectance of the orchid flowers is very similar to that of *D. ulicifolia*, and analysis of spectra in a vision model suggested that bees would have difficulty discriminating between the orchid and the legume species on the basis of flower color alone (Indsto et al. 2006). Furthermore, both the orchid and the legumes have similar UV-absorbing nectar guides. Earlier studies showed that *D. maculata* shares pollinators with a number of similar "egg and bacon" legume species (Beardsell et al. 1986), making this a likely case of guild mimicry, rather than mimicry of a specific species.

The functional significance of floral color patterns in the food-deceptive Asian orchid *Paphiopedilum micranthum* has been studied in Guizhou Province in southwestern China (Ma et al. 2016). This orchid exhibits a striking color contrast between its pink–white petals (bee blue) and central UV-absorbing yellow (bee green) staminode. Bumblebees foraging on pollen-rewarding flowers with a similar color contrast between the petals and anthers are more likely to visit the orchid than are those feeding on flowers with dissimilar color patterns. Ma et al. (2016) found that the probability of visitation by bumblebees decreased with color distance (in a bee vision model) between the orchid and food flowers for both outer and inner floral parts. In the communities in which the orchid grows, species of *Rosa* (Rosaceae) that share the contrasting color pattern with the orchid also shared its assemblage of bumblebee pollinators, while plants with different floral color patterns had pollinator assemblages dissimilar to those of the orchid. It is not known whether *P. micranthum* qualifies as a Batesian mimic of any specific pollen-rewarding species, but it does seem that the color contrast in flowers of this species is important for the exploitation of relationships between bumblebees and their pollen flowers. A phylogenetic analysis of *Paphiopedilum* showed that the colors of the outer and inner parts of *P. micranthum* flowers have different evolutionary histories (Ma et al. 2016). The pink perianth is a novel (apomorphic) trait and thus likely to be an adaptation for attracting bumblebees, while the UV-absorbing yellow color of the staminode is ancestral and probably evolved long before its current function of exploiting bumblebees.

Further evidence for the functional significance of the color pattern of *P. micranthum* for attracting pollen-seeking bumblebees was obtained from experiments in which the UV-absorbing yellow staminode was excised from some flowers and left intact in control flowers (Ma et al. 2016). Flowers lacking the central staminode experienced much lower levels of pollen removal than did the intact control flowers. Bees tended to closely inspect flowers in which the staminode was replaced by a UV-absorbing yellow paper disk, but not flowers with UV-reflecting yellow or green disks. However, it is uncertain whether this behavior is a generalized response of bees to signals that imitate pollen (Lunau 2000) or whether it has been shaped by experience with flowers in the local community.

Flower shape and inflorescence architecture

The ability of pollinators to detect fine details in flower shape is probably limited, except at very close range (Vorobyev et al. 1997; Yoshioka et al. 2007). While scent and color are critical for attracting pollinators from long and intermediate distances, respectively, flower shape can be important for close-up decisions by pollinators about landing. For example, one study revealed that plastic flowers with a dissected outline were not only inspected more often by long-proboscid flies under field conditions than were those with a simple conical shape (Figs 4.4 and 4.5), but they were also much more likely to be probed (Jersáková et al. 2012). This does not mean that mimics need to resemble the models in every aspect of flower shape. It has been noted, for example, that pollinators will often probe flowers that are heavily damaged by herbivores, suggesting that fine details of flower outline are not necessarily critical for eliciting flower-probing behavior (Johnson and Morita 2006).

A potential example of shape mimicry occurs in some Neotropical orchids belonging to the Oncidiinae (Fig. 4.2). The Malphigiaceae that appear to serve as models for these orchids have a distinctive, deeply dissected floral outline made up of five claw-like petals. The orchids have a similar floral outline due to dissection of the labellum and modifications of the size and orientation of the remaining flower parts. Using eigenshape morphometrics, Papadopulos and his co-workers (Papadopulos et al. 2013) showed that the flowers of some orchids in the Oncidiinae are more similar in shape to those of Malphigiaceae than they are to those of other plant species in the same communities. The novelty of the malphig-like shape among these orchids makes it likely that it has a mimicry function. Unfortunately, we are not aware of any field experiments that have tested the functional significance of flower shape or even color for the attraction of oil bees to orchids in the Oncidiinae. This huge system consisting of hundreds of species of orchids and vines in the Neotropics represents one of the great unexplored frontiers of floral mimicry research.

There is some evidence from a different mimicry system that pollinators can use inflorescence architecture as an attraction cue. In a study of two orchids that mimic the rewarding species *Scabiosa columbaria*, long-proboscid flies were not attracted to inflorescences of one of the mimics that had been artificially reconstructed in the shape of a spike rather than the flat-topped capitulum that is typical of the model and mimics in this system (Johnson et al. 2003a) (see Fig. 7.3). However, in other experiments with different species of long-proboscid flies, inflorescence shape (raceme versus umbel) had no effect on attraction (Jersáková et al. 2012), suggesting that inflorescence shape is not a universal signal for pollinator attraction, even among the same group of flower visitors.

Floral scent

Many authors have reported that, to the human nose, the flowers of Batesian mimics are scented weakly or not at all (Beardsell et al. 1986; Johnson 1994, 2000). Actual quantification of scent emissions has, however, rarely been attempted.

The few available data suggest that pollinators can be deceived even when mimics and models differ in scent. For example, bees regularly move between the rewarding flowers of *Bellevalia flexuosa* and the food-deceptive orchid mimic *Orchis israelitica* (Dafni and Ivri 1981), suggesting that they are regularly duped. The two species have very similar flower colors, and dark spots on the orchid labellum even resemble the dark anthers of *Bellevalia*. The loci of reflectance spectra of the orchid flowers are much closer to those of *Bellevalia* flowers in a bee vision model than are those of related

orchids. Yet the scent emissions of these two species are quite different in overall composition and emission rates (the orchid has a much weaker scent). Bee pollinators would be likely to perceive these scents as different, as evidenced by calcium imaging studies of the location of responses in brains of honey bees exposed to the scent of the two species (Galizia et al. 2005). One important caveat to this study is that the honey bees used for calcium imaging are not the primary pollinators of *O. israelitica*. However, it does seem evident that bees approach and probe flowers of the orchid because of its visual similarity to the *Bellevalia* flowers and are not deterred by the mismatch in odors. Odors may play a role in long-distance attraction, but since mimics and models are often intermingled, differences in odor among plants may not significantly affect alighting decisions.

Thelymitra orchids, proposed as mimics of pollen-rewarding flowers, are often strongly scented. *Thelymitra macrophylla*, for example, produces a scent which is stronger than and chemically very different from that produced by its putative model, *Orthrosanthus laxus* (Iridaceae) (Edens-Meier et al. 2014). The significance of this difference in odors between the two species is difficult to gauge, as there is no substantial evidence for pollinator sharing by these two species, nor tests of pollinator discrimination between the species in the field.

Interactions between oligolectic (specialized) bees and their food plants are often mediated by scent (Dötterl et al. 2005; Milet-Pinheiro et al. 2013). A non-rewarding plant that exploits these interactions may need to mimic the key volatile signals that are used by the bees to locate their food plants. Nilsson (1983b) showed that the food-deceptive orchid *Cephalanthera rubra* exploits the interaction between *Chelostoma* bees and *Campanula* species (Campanulaceae). Many of the bees found sheltering in *Campanula* flowers carried pollinia of the orchid, and the lilac colors of the orchid and *Campanula* flowers were found to be closely matched in the spectral range significant to bees. Recent studies have shown that the initial attraction of *Chelostoma rapunculi* bees to *Campanula* flowers is mediated both by color and spiroacetals, a relatively rare class of volatiles, and that newly emerged bees rely on these compounds to locate host flowers

(Milet-Pinheiro et al. 2013). It is thus possible that *C. rubra* emits these compounds in order to exploit the interaction between *Chelostoma* and *Campanula*. Visual cues alone can elicit landing responses in *Chelostoma* bees (Milet-Pinheiro et al. 2012), and even inexperienced *Chelostoma* bees have a preference for blue colors similar to those of its food plants (Milet-Pinheiro et al. 2015), further calling into question the assumption that *C. rubra* relies primarily on conditioned bees for pollination. Another piece of evidence that suggests that *C. rubra* may exploit innate behavior is that it often flowers earlier than *Campanula* and is pollinated by early emerging male *Chelostoma* bees (Nilsson 1983b).

Another likely food mimicry system in which scent may play a key role is that between orchids in the Oncidiinae and oil-producing Malphigiaceae. It is now well known that scent plays a key role in the location of food plants by oil-collecting bees (Schäffler et al. 2015). The oil-derived volatile diacetin is used by bees as a cue to locate oil-producing flowers and this volatile has been detected in extracts from many oil-producing plants, including Malphigiaceae and some of the oil-producing species in the Oncidiinae (Schäffler et al. 2015). Diacetin is likely to be a reliable cue for the location of oil-producing flowers as it is produced by the same biochemical pathway that leads to oil production. If so, an intriguing question is how various Oncidiinae that are considered to be food deceptive attract their bee pollinators. Could imitation of the visual signals of Malphigiaceae be sufficient, or do Oncidiinae that resemble malphigs always produce small amounts of oil along with diacetin, or even produce diacetin and other oil-associated signals via a different pathway?

In one recently discovered food-source mimicry system the mimic appears to rely almost entirely on scent signals to attract pollinators. The trap flowers of the Mediterranean pipevine, *Aristolochia rotunda* (Aristolochiaceae), are pollinated primarily by small flies of the family Chloropidae. These flies are kleptoparasites that feed on the victims of arthropod predators. The chloropid flies pollinating *A. rotunda* are extraordinarily specialized for feeding on the exudates of freshly killed bugs (Miridae), and use the volatile compounds that are emitted when bugs are dismembered by arthropod predators to locate their food. The flowers of *A. rotunda* emit a set of volatile compounds (mainly aliphatic esters) that are similar in chemical composition to those emitted by freshly killed bugs, thereby deceiving their chloropid fly pollinators (Oelschlägel et al. 2015). Other kleptoparasitic fly families (e.g., Milichilidae) have also been recorded as flower visitors. Indeed, chloropids and milichilids are important pollinators in another group of plants with trap flowers, namely the *Ceropegia* milkweeds (Ollerton et al. 2009; Heiduk et al. 2015). Studies of the Asian species *Ceropegia dolichophylla* showed that spiroacetals emitted by the flowers elicit behavioral responses in kleptoparasitic flies (Heiduk et al. 2015). Although the details of this mimicry system are not completely understood, it is known that spiroacetals are emitted by various insects. It is possible that floral mimicry of arthropod food sources for flies and wasps is actually a widespread form of food-source mimicry among plants (Bower et al. 2015). We discuss wasp pollination of flowers that smell like prey in Chapter 7, because these systems involve brood provisioning (Brodmann et al. 2008, 2009).

Pollinator discrimination between mimics and models

It is not an absolute requirement of Batesian mimicry that an operator should be unable to discriminate between mimics and models, only that it should have difficulty in doing so, thus enhancing the fitness of the mimic (Ruxton et al. 2004). Even if operators possess the ability to discriminate between mimics and models, they do not necessarily do so in nature because discrimination uses valuable time and making a few mistakes while foraging rapidly may ultimately be less costly to the operator than slow and deliberate foraging (Chittka et al. 2009). It follows that mimics need not be carbon copies of the models, only that they should sufficiently resemble their models such that operators make mistakes in discrimination. Therefore, observations of the behavioral responses of operators to mimics and models is ultimately a much better test of their similarity than is the direct measurement of signals, with all of the problems associated with interpretation given the differences in how humans perceive signals compared with other animals.

Field measurements of the ability of operators to discriminate between mimics and models are now available for several plant mimicry systems. Indeed, operator discrimination is one of the aspects of research on Batesian mimicry that is generally more tractable for plant systems than it is for animal systems. Early attempts to measure the discrimination ability of pollinators involved observations of mixed patches of mimics and models and assessment of whether the interplant movements of pollinators conformed to a random null model (Johnson 1994). Later methods included the use of a portable "presentation stick" to offer a choice of model and mimic to foraging pollinators (Box 5.2) (Johnson 2000; Johnson et al. 2003a; Anderson et al. 2005). These bioassays have yielded a striking result: in most cases, pollinators do not appear to discriminate between the mimics and models, despite the lack of rewards in flowers of the former and the lack of precise matching in floral phenotypes.

There are three possible interpretations for the apparent lack of discrimination by pollinators in many food-source mimicry systems. The first is that pollinators are generalist and non-discriminating flower visitors. This possibility can be rejected out of hand since the pollinators in these systems are nearly always highly specialized foragers that visit a small and select number of plant species in each community (Anderson et al. 2005). The second possibility is that pollinators are able to discriminate but fail to do so in order to avoid costs in terms of discrimination time (Chittka et al. 2009). Pollinators nearly always leave the mimics after probing a single flower, implying that they could reduce costs by using a system of discrimination based on the taste of the first one or two flowers, rather than a remote one based on signal detection (Johnson et al. 2004). The third possibility is that pollinators find it very difficult or even impossible to discriminate remotely between mimics and models (Benitez-Vieyra et al. 2007; Peter and Johnson 2008). One way to test whether pollinators do not discriminate because of speed–accuracy trade-offs or because they are incapable of doing so is to increase the cost of mistakes by, for example, adding bitter toxins to the flowers of the mimics. If this results in avoidance, one could conclude that the

lack of discrimination in the field is simply because of speed–accuracy trade-offs.

An interesting set of glasshouse experiments conducted by Gigord et al. (2002) highlighted the potential role of pollinator conditioning in Batesian food-source mimicry systems. These experiments involved the European deceptive orchid *Dactylorhiza sambucina*, which has both red and yellow color morphs and is pollinated through a system of generalized food deception (see Chapter 3). As may be expected, naïve bumblebees did not discriminate between the two color morphs. However, bumblebees that had obtained nectar from the red morph of another plant species, *Mimulus gutattus* (Phrymaceae), preferentially visited the red orchid morph, while those that had obtained nectar from a yellow morph of *M. guttatus* preferentially visited the yellow orchid morph. If the orchids and the rewarding *Mimulus* plants were kept together, bees increasingly and eventually exclusively preferred the latter, showing that they were easily able to discriminate between the species, presumably by using shape and scent cues. This is quite different from the situation in most Batesian food-source mimicry systems where pollinators consistently fail to discriminate between mimics and models (Johnson 2000; Johnson et al. 2003a). This experiment shows that a superficial color resemblance is not sufficient to prevent eventual discrimination by social bees that alternately encounter mimics and models in the field. However, it also shows that associative learning of color traits is likely to be an important component of Batesian food-source mimicry systems.

Although lack of pollinator discrimination between mimics and models is consistent with mimicry, there is no reason why mimics should not benefit even if pollinators are capable of some degree of discrimination. For example, mimics may have floral displays that provide similar color signals to those of their models, but may be different in shape, or smaller. A good case in point is the orange-flowered form of the orchid *Disa ferruginea* (cluster orchid) which mimics an orange-flowered asphodel (*Kniphofia uvaria*) in the Southern Cape region of South Africa (Plate 2b, c). Mountain pride butterflies (*Aeropetes tulbaghia*) exhibit a preference for the asphodel over the cluster orchid (de Jager et al. 2016), probably because the former tends to have inflorescences that are larger and have more flowers than those of the orchid, but the orchid nevertheless regularly receives mistake visits and is pollinated by the butterflies. Interestingly, the same butterfly species seems to find it harder to discriminate between a red-flowered form of the cluster orchid and its red iris model (*Tritoniopsis triticea*) in the Western Cape of South Africa (Johnson 1994), perhaps because the red-flowered form of the orchid is a more accurate mimic of the inflorescence size and shape of this particular model. Therefore, some degree of pollinator discrimination between plant species cannot be used to refute a hypothesis that one is a mimic of the other. Indeed, since there is likely a continuum between Batesian floral food-source mimicry and generalized food deception, we can expect cases in which mimicry is imperfect but nevertheless functional. Finally, it should also be borne in mind that some sexually deceptive mimics have been shown to produce super-normal stimuli that actually render them more attractive than their models (Vereecken and Schiestl 2008). We are not aware of such super-normal stimuli in food-source mimicry, but they are a theoretical possibility, particularly in scent-based systems involving innate attraction.

Evidence that mimics benefit from the presence of models

A basic prediction of Batesian mimicry is that the fitness of the mimic should be enhanced by the presence of its model(s). The underlying assumption for this prediction is that mimics should benefit when the operator is conditioned by experience with the model. While it may seem simple to compare the performance of putative mimics between sites where models are present and sites where models are absent, there are some major challenges in the interpretation of results of such experiments. In the case of putative food-deceptive plant mimics that benefit from the presence of certain rewarding plants, it is very difficult to establish if this is due to the conditioning of pollinators on the signals of the rewarding plants or because pollinators are simply more abundant in the vicinity of the rewarding plants. The latter phenomenon is known as the magnet-species effect and describes the situation when certain highly attractive rewarding plants

result in local aggregations of pollinators that benefit other, less attractive species (Laverty 1992). Orchids that deploy generalized food deception and even some rare rewarding plant species can benefit from the magnet-species effect even though they are not specific mimics of the rewarding magnet species (Johnson et al. 2003b).

The issue of ecological facilitation of pollination in mimics by models was the subject of one of the first major controversies in plant mimicry research. Thomas Boyden (1980) suggested that the deceptive orchid *Epidendrum radicans* might be a mimic of the rewarding flowers of *Asclepias curassavica* and *Lantana camara*. All three species have similar displays of orange and yellow flowers. According to the tropical-orchid biologist Calaway Dodson, the idea of mimicry between these species had originated during a course run by the Organization for Tropical Studies (OTS) in 1966 and had become a regular subject for small field studies conducted by students (Bierzychudek 1981). All three species are pollinated by butterflies, and Boyden had collected butterflies in Panama that were carrying *Epidendrum* pollinia on their proboscides, *Asclepias* pollinia on their legs, and the granular pollen of *Lantana* on their heads. What seemed at first to be an excellent example of floral mimicry fell apart when Bierzychudek (1981) pointed out two major problems with Boyden's original mimicry hypothesis. Firstly, there is uncertainty whether the "models" *L. camara* and *A. curassavica*, which colonize disturbed areas, are actually native to Central America. Secondly, a small data set that Bierzychudek collected in Costa Rica in the 1970s showed that the rate of removal of pollinaria from flowers of *Epidendrum* was unrelated to the presence of *Asclepias*. Follow-up studies in Brazil (Fuhro et al. 2010) have confirmed extensive sharing of butterfly pollinators among *A. curassavica*, *L. camara*, and a similar orchid species *Epidendrum fulgens*, but whether or not these *Epidendrum* species are Batesian mimics or simply rely on generalized food deception remains unknown.

Ecological facilitation seems to be important in other potential mimicry systems. Dafni and Ivri (1981), for example, reported that fruit set in *Orchis israelitica* was several-fold greater at sites where it occurred with its proposed model *Bellevalia flexuosa*. Peter and Johnson (2008) similarly found that the

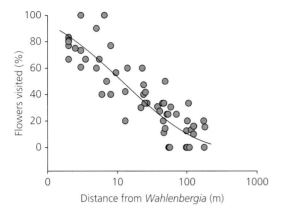

Figure 4.6 Flowers of the orchid mimic *Eulophia zeyheriana* are more likely to be visited by pollinators when they occur in close proximity to the rewarding model *Wahlenbergia cuspidata*. Adapted from Peter and Johnson (2008).

proportion of flowers of *Eulophia zeyheriana* visited by pollinators declined systematically with distance away from clumps of the rewarding model *Wahlenbergia cuspidata* (Fig. 4.6). This result could be replicated when orchids were physically translocated either into or 40 m away from patches of *W. cuspidata*, indicating a strong pollination benefit of close co-occurrence of the mimics with their models.

Researchers in Mexico found that fruit set in the rewardless orchid *Oncidium cosymephorum* is enhanced by the presence of the oil-producing shrub *Malphigia glabra* (Malphigiaceae) (Carmona-Diaz and Garcia-Franco 2009). This shrub has flowers that are remarkably similar in color and shape to those of the orchid and shares pollinators, namely *Centris* oil-collecting bees, with the orchid. Orchids isolated from *Malphigia* shrubs frequently did not produce any fruit at all. Most other studies of likely Batesian food-source mimicry indicate that mimics benefit from the presence of models, but achieve some pollination success when isolated from models.

It is not a strict prediction that Batesian food-source mimics will depend on the presence of models to attract pollinators, only that the presence of models will enhance pollination success. In the absence of models, Batesian food-source mimics will usually attract some visits by pollinators, either because the pollinators have been conditioned at distant sites or because the

flowers have some functionality in terms of generalized food deception.

Frequency dependence

One of the key theoretical predictions of Batesian mimicry is that the fitness of mimics should decline as they become more numerous relative to their models (Fisher 1930). An ancillary prediction is that very high frequencies of mimics may have a negative impact on the fitness of models. These predictions of frequency dependence are based on the underlying assumption that Batesian mimicry depends on a learning process (Fisher 1930). In reality, frequency dependence has rarely been examined in wild populations, either for animal or plant mimicry systems. Most of the studies of frequency dependence in Batesian protective mimicry have involved laboratory-type experiments, such as manipulations of the relative abundance of distasteful and palatable food items available to domesticated chickens (Ruxton et al. 2004). This reflects the difficulties in measuring and manipulating the relative frequencies of animal mimics and models in the field. It is also not straightforward to attribute the fitness of mimics to the relative frequencies of mimics and models at different sites because the abundance of operators (predators or pollinators) typically also varies between sites.

In a study conducted in natural habitats in the Drakensberg mountains of South Africa, Anderson and Johnson (2006) used experimental arrays of an orchid mimic (*Disa nivea*) and its model (*Zaluzianskya microsiphon*) to test the predictions of frequency-dependent fitness. As predicted, the fitness of the orchids, as estimated by rates of pollen deposition and removal, declined sharply when mimics were frequent relative to their models in small experimental patches of five plants (Fig. 4.7). Flies did not discriminate between mimics and models, but visits to orchids resulted in fewer flowers being probed and longer subsequent flight distances compared with visits to the model. Thus, the results could be explained by flies moving out of patches that contained a high frequency of orchids. There was no evidence in this study for the conventional idea that operators in mimicry systems learn to avoid the signals of mimics when they are common. It also

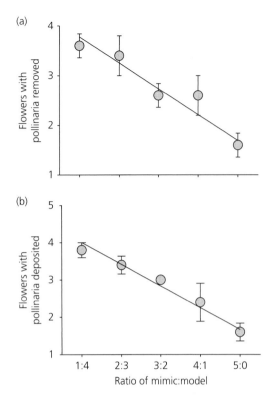

Figure 4.7 Effects of the relative numbers of non-rewarding mimics and rewarding models on male (a) and female (b) components of pollination success in the orchid mimic *Disa nivea*. Adapted from Anderson and Johnson (2006).

seemed unlikely that pollinators use a learned mental map to avoid patches that have a high frequency of mimics, as shown in some studies of sexually deceptive mimics (Wong and Schiestl 2002). The density dependence in the *Disa–Zaluzianskya* case seemed to arise because pollinators quickly move out of patches that contain a high frequency of non-rewarding orchids and, conversely, forage for longer in patches that contain a high frequency of rewarding models. Interestingly, fitness of the *Zaluzianskya* model plants in this study was not affected by the relative frequency of mimics and this was attributed to the tendency of the pollinator to probe many flowers of the model per visit, so that fewer visits did not automatically translate into lower fecundity.

The general prediction that Batesian mimics should be less abundant than their models is based on the assumption that the mimic to model ratio will influence the behavior of operators through learning (Ruxton et al. 2004). Negative frequency dependence (where mimics perform best when they are rare relative to models) may arise because operators learn to discriminate between models and mimics when encounters with mimics become frequent. The absolute number of models, on the other hand, may also shape operators' responses to the model phenotype and thereby indirectly benefit mimics. Because pollinators rarely seem to discriminate between Batesian food-source mimics and models, it may be that the relative frequency of mimics and models is not as important as was previously believed.

The geography of Batesian food-source mimicry

Nested distributions

Ecological facilitation of the performance of mimics by models should lead to the distributions of mimics being nested within those of the models. Similarly, species that have adapted to similar habitats would be more likely to subsequently develop mimic–model relationships. Of course, mimics are not necessarily completely ecologically dependent on models, and a species may mimic several different models across its distribution range. Nevertheless, there is now good evidence for associations between the geographic distributions of mimics and their models. These range from continental-level associations, such as that between the Oncidiinae and the Malphigiaceae that are both concentrated in the Neotropics, and much finer-scale associations between the distributions of species that appear to have a single model. The most detailed studies of the distributions of plant mimics and their models have come from South Africa (Johnson 1994, 2000; Anderson et al. 2005; Peter and Johnson 2008) (Fig. 4.8). The open landscapes of South Africa make detailed distribution mapping far more feasible than in tropical rainforests, where distribution information is patchy and often limited to collections from roadsides. In addition, the Batesian food-source mimicry systems described from South Africa are often highly specialized and frequently

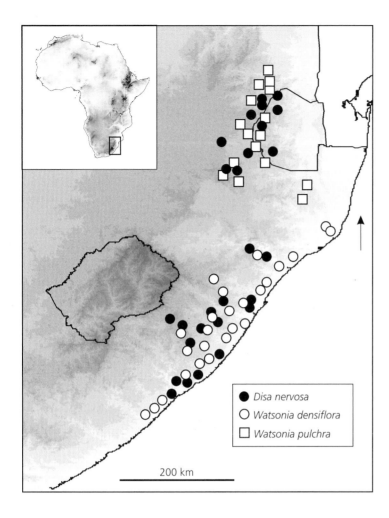

Figure 4.8 The geographic association between the food-deceptive orchid *Disa nervosa* and its two food-rewarding models, *Watsonia densiflora* and *Watsonia pulchra*. The tabanid fly that pollinates these plants has a much wider distribution across southern Africa. Adapted from Johnson and Morita (2006).

involve plant species pairs, making them ideal for testing some of the biogeographic predictions of the mimicry hypothesis.

Arguments about associations between the distributions of food-source mimics and their models are much more compelling when the pollinator has a wider distribution than the mimics or even the models. In such cases, the restricted distribution of the mimic within that of its model cannot be explained by a lack of availability of suitable pollinators outside the range of the mimic. Indeed, in most of the food-source mimicry systems that have been studied, pollinator distributions are much broader than those of the floral mimics. A major challenge for future research is to determine whether the distributions of mimics are restricted by a lack of

sufficient pollination outside the range of the models, or by physiological adaptations to particular environments. Of course, both explanations may apply, as in the plausible scenario of species that are physiologically limited to particular physical habitats and subsequently develop adaptations to the particular plant communities in those habitats, including mimicry.

Trait tracking

Over time mimics will have to undergo evolutionary responses to any changes in the signals of their models. This is an example of the Leigh van Valen's (1973) evolutionary analogy of the Red Queen, based on Lewis Caroll's fictional character who stated that it was necessary to keep running merely to stay in the same place. Since the evolution of mimics is advergent (one-sided) with respect to their models, this process can be described as trait tracking. If the model and its pollinators are engaged in a coevolutionary arms race, as is likely for many long-tubed flowers and long-proboscid animals (Thompson 1994), then the mimic will have to track the trait evolution that results from this coevolutionary engine of model and operator. It is not easy to reconstruct the evolutionary changes that mimics have had to undergo in response to changes in their models over time, but we can use a space-for-time substitution to determine how the process may work.

The South African food-deceptive orchid *Disa nivea* has been proposed to be a Batesian mimic of the co-occurring and nectar-rewarding species *Zaluzianskya microsiphon* (Anderson et al. 2005). The two species have closely matched flower colors and share the same long-proboscid fly pollinator, which does not discriminate between the two species under field conditions. There is a significant co-association between the size of the flowers (a likely visual signal) of the orchid and its model over a wide geographic region in the Drakensberg mountains (Anderson et al. 2005) (Fig. 4.9a). There is also an association between the flower depths of the two species across sites (Fig. 4.9b). It seems likely that the length of the fly mouthparts and the depth of the flower tube in the *Zaluzianskya* species have coevolved, while the *Disa* species (which is non-rewarding and therefore cannot influence the evolution of fly proboscis length) has unilaterally tracked the evolutionary changes in the mouthparts of the fly pollinator. Therefore, in this system, the orchid may have tracked the evolution of both flower size in its model and proboscis length in its pollinator. However, since there are probably some developmental allometric correlations between flower width and spur length in the orchid, further experiments based on floral manipulation would be required to confirm the direct targets of selection on floral traits of the orchid.

Geographic polymorphism

One of the most distinctive features of Batesian mimicry is the evolution of geographic polymorphism

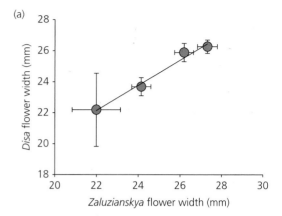

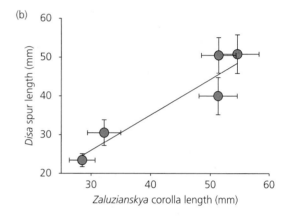

Figure 4.9 Co-variation of floral trait dimensions among populations of the food-deceptive orchid *Disa nivea* and its rewarding model *Zaluzianskya microsiphon*. (a) Flower width. (b) Flower depth. Adapted from Anderson et al. (2005).

in color patterns. This is well known in a wide range of animal groups including butterflies (Vane-Wright et al. 1999), salamanders (Brodie and Brodie 1980), and snakes (Greene and McDiarmid 1981). These patterns are attributed to shifts between different models and it reflects an important process that could drive adaptive diversification in mimics. Does the same phenomenon occur in floral food-source mimicry?

According to Bierzychudek (1981), the legendary tropical-orchid biologist Calaway Dodson once suggested that the striking geographic variation in the flower color of the food-deceptive orchid *Epidendrum secundum*, which ranged from yellow to cerise to pink along a north–south

transect of the Andes, could be explained by
the prevailing flower color of sympatric *Lantana*
(Verbenaceae) species. To our knowledge, Dodson's hypothesis of model-driven geographic
color polymorphism in *Epidendrum* has not been
investigated.

The best-studied example of geographic polymorphism in a floral mimicry system involves
the South African orchid *Disa ferruginea*. The geographically separated red and orange color forms
of this butterfly-pollinated species have been ascribed to mimicry of different nectar-producing
model species (Johnson 1994). The model for the
red form of the orchid in the western Cape region
appears to be a red-flowered iris (*Tritoniopsis triticea*), while that for the orange form of the orchid
in the southern Cape region is the orange-flowered
asphodel (*Kniphofia uvaria*). Reflectance spectra of
flowers of the two forms of this orchid match those
of its putative models (Fig. 4.10a). To test the idea
that color forms of the mimic are locally adapted
to different models, a team of biologists conducted
reciprocal translocation experiments with the orchids (Newman et al. 2012). In addition, they used
model flowers to control for possible shape or odor
cues. The results showed clearly that the butterflies
prefer forms with colors that match those of the
flowers of local nectar plants (Fig. 4.10b). More specifically, red was favored over orange at the sites
where the butterfly obtains nectar from red flowers
and orange was favored over red at sites where the
butterfly obtains nectar from orange flowers. Since
both red and orange forms of the orchid are pollinated by the same butterfly species (*Aeropetes tulbaghia*), associative conditioning on flower signals
from locally co-occurring models is the most likely
basis for selection on flower color in the orchid
mimics. The alternative possibility, that the butterfly has developed different innate color preferences
in different parts of its range, cannot be ruled out
in this case.

Taxonomic bias in the pollinators of food-source mimics

Non-rewarding plants that rely on generalized food
deception tend to be pollinated by a range of different pollinators, which in turn tend to be relatively

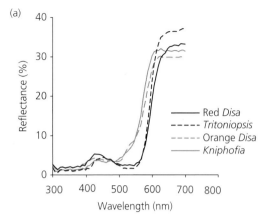

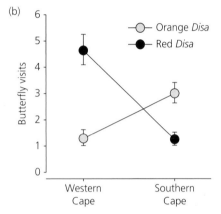

Figure 4.10 Color matching between the food-source mimic *Disa
ferruginea* and its model plants is consistent with local adaptation.
(a) Floral spectral reflectance of the red-flowered western form of
D. ferruginea matches that of *Tritoniopsis triticea*, while spectral
reflectance of the orange-flowered southern form of *D. ferruginea*
matches that of *Kniphofia uvaria*. (b) Responses of the butterfly
Aeropetes tulbaghia to reciprocally translocated orchid inflorescences.
The red color form of *D. ferruginea* is most attractive to butterflies
in the western Cape where butterflies feed from red flowers, but the
orange form is most attractive to butterflies in the southern Cape
where butterflies feed from orange flowers. Adapted from Newman
et al. (2012).

generalized foragers. These pollinators, such as
bumblebees, continuously explore potential new
food sources and can easily be duped by generalized food deception. Batesian food-source mimics,
by contrast, tend to exploit relatively specialized
mutualisms between plants and pollinators, with
both the mimics and the pollinators in these systems tending to show a high level of specificity

(Anderson et al. 2005). In these specialized systems, pollinators are seeking very specific signals from their food plants within the community of flowering plants; a mimic must match these signals closely in order to attract the attention of the specialist pollinator (Jersáková et al. 2012).

As is the case for many other floral mimicry systems (see Chapter 5), the taxonomic composition of pollinators involved in Batesian food-source mimicry usually differs markedly from the regional pool of pollinators. In South Africa, for example, most of the insect species found to be involved in Batesian food-source mimicry are long-proboscid flies (Nemestrinidae or Tabanidae) or butterflies (Johnson 1994, 2000; Anderson et al. 2005), even though these insects make up only a small fraction of the local pollinator pool. Most of the proposed Batesian food-source mimicry systems in European and Australian plants involve solitary bees (Dafni and Ivri 1981; Nilsson 1983b; Beardsell et al. 1986), which is less surprising given that these insects generally play an important role in plant pollination. There are also several proposed cases of Batesian food-source mimicry involving bumblebees (Kjellsson et al. 1985; Sugiura et al. 2002), but these require further study to determine if the floral adaptations are specific to the interactions with a particular model.

Pollinator groups that are strikingly underrepresented in floral Batesian mimicry systems worldwide are birds, moths, beetles, and social bees. There may be differing reasons for the underrepresentation of these groups; birds, for example, may discriminate against mimics using visual cues, given their acute vision systems, while beetles, moths, and social bees may rely heavily on olfaction when making flower choices. While mimicry of scent cues to attract these insects would be possible, it would require the evolution of a scent blend which is close enough to that of the model to result in cognitive misclassification.

It has been suggested that the insects most easily deceived would be those that use mainly color cues for foraging (Jersáková et al. 2006a). This seems to be the case for the long-proboscid flies in the *Disa pulchra* mimicry system. These flies do not discriminate between completely unscented artificial flowers and real flowers, as long as the spectral reflectance is matched (Fig. 4.5).

Evidence for the idea that flies are more easily duped by floral mimics than are bees was recently obtained by Jersáková et al. (2016). They studied the European food-deceptive orchid *Traunsteinera globosa*, which is considered to mimic a guild of nectar-producing plants with flat-topped inflorescences belonging to the genera *Knautia* and *Scabiosa* (Dipsacaceae) and *Valeriana* (Caprifoliaceae). The compressed inflorescence of this orchid is a novel feature in its lineage (and thus likely a mimicry trait) and the flowers match those of its model guild in terms of spectral reflectance, but not floral scent (Jersáková et al. 2016). When inflorescences of *T. globosa* were proffered to insects foraging on flowerheads of the various model species, hoverflies (Syrphidae) and dance flies (Empididae) were readily duped and probed the orchid, but bees tended to inspect the orchid and then fly away without probing. It is not known whether bees use visual or scent cues, or both, to discriminate against the orchid. The evolution of floral mimicry may be more likely to occur if it requires only modification of fewer signaling modalities, for example only color and shape, but not scent.

Guild mimicry

The continuum between generalized food deception and Batesian food-source mimicry ranges from exploitation of opportunistic foraging behavior through to exploitation of behavior that is associated with a very specific model phenotype. Along this continuum lie several examples of food-deceptive plants which engage in more generalized forms of mimicry, typically exploiting the interactions between pollinators and entire guilds of plants. The globe orchid *Traunsteinera globosa*, discussed above, is an excellent example of guild mimicry. Similarly, *Diuris* and *Thelymitra* orchids in Australia appear to exploit the interactions between bees and a range of food plants. The signals of these orchids cannot be ascribed to mimicry of any one specific food plant. The interactions between Oncidiinae and Malphigiaceae in the Neotropics are probably also not specialized at the species level and are best described as guild mimicry. Floral traits of the Asian orchid *Paphiopedilum micranthum* appear to be adaptations for exploiting an entire guild of plants with flowers that are used by bumblebees as sources of pollen (Ma et al. 2016).

Batesian food-source mimics that exploit behavior associated with a specific floral phenotype should have a strong ecological dependence on the particular model species and are usually accurate mimics. Guild mimics, on the other hand, do not have an ecological dependence on any one model species and tend to be imperfect mimics. One of the explanations for imperfect mimicry in animals is adaptation to multiple models (Edmunds 2000) and we suggest that this also applies to most examples of guild mimicry in food-deceptive systems in plants. By comparison, imperfect mimicry in sexual and oviposition-site mimics is more likely to be explained by receiver bias (see Chapters 5 and 6).

Overview and perspectives

Batesian food-source mimicry in plants has proved to be tractable to a number of lines of investigation that are difficult to accomplish using animal subjects. The most rewarding studies have involved direct experimental studies of operator discrimination between mimics and models, followed up by manipulations of the traits of mimics to determine which are responsible for the cognitive misclassification by operators. Geographic polymorphism in Batesian food-source mimics also provides an excellent opportunity to use reciprocal translocations to test whether forms are adapted to the local environment of model organisms.

The most glaring gaps in our knowledge about the evolution of Batesian food-source mimicry relate to the role of pollinator conditioning. We know that these mimics exploit specialized mutualisms between plants and pollinators, but are the pollinators in these mutualisms hard-wired for preferring particular signals or do they show increasing preference for these signals as they encounter rewarding plants? The answer to this question is critical for understanding the frequency dependence of mimics and models, and also the broader question of whether Batesian food-source mimicry is something entirely distinct from floral syndromes, or just a special case involving advergent evolution.

CHAPTER 5

Sexual mimicry

Introduction

The resemblance of some flowers to insects has long fascinated human observers, including Linnaeus, who coined the name *Ophrys insectifera* for a European orchid species in 1753. It was only some 160 years later, however, that the true meaning of insect resemblance in flowers was discovered. The first correct interpretation of sexual mimicry was provided by the French amateur botanist Maurice-Alexandre Pouyanne, who had observed the pollination of *Ophrys speculum* in Algeria over many years (Pouyanne 1917). He suggested that male scoliid wasps that alight on the flowers of this species mistake them for female scoliid wasps, and hence attempt to copulate with them. Pouyanne's 20 years of observations and his interpretations were later confirmed by Godfery (1925) who observed sexual mimicry in other *Ophrys* species, and by Edith Coleman who discovered sexual mimicry in the Australian genus *Cryptostylis* (Coleman 1927).

The Swedish entomologist Bertil Kullenberg (Fig. 5.1) almost single-handedly developed the nascent program of research on sexual mimicry and conducted an extensive series of experiments from the 1950s to the 1990s, some of the early results of which he published in 1961 in a seminal monograph (Kullenberg 1961). Kullenberg's work established the basic principles of sexual mimicry. In essence, he showed that floral sexual mimics imitate the mating signals of receptive female insects. These signals consist of chemical, visual, and tactile components which have a different significance in different systems, chemical mimicry usually being of primary importance. Kullenberg showed this through careful experiments in which he eliminated either visual or chemical signals and tested the

attractiveness of such manipulated flowers to their pollinators.

Kullenberg's ecological experiments inspired a new generation of researchers devoted to the identification of the volatile compounds emitted by *Ophrys* flowers (Bergström 1978; Borg-Karlson 1990). In the meantime, the Austrian evolutionary ecologist Hannes Paulus and the Australian entomologist Colin Bower independently established that sexual mimics are often pollinated on a highly specific, even species-specific basis, which has important consequences for reproductive isolation (Paulus and Gack 1990; Bower 1996). Both researchers found that floral volatiles form the basis of the specific pollinator attraction. However, identification of the volatile compounds that elicit copulatory behavior in insects proved to be more difficult than anticipated. The key breakthrough in the identification of these chemicals occurred only at the end of the twentieth century, when behaviorally active volatiles were identified in both a sexual mimic (*Ophrys sphegodes*) and its model, the solitary bee *Andrena nigroaenea* (Schiestl et al. 1999). This study involved the use of gas chromatography combined with electroantennographic detection (Box 5.1), and showed that the compounds attractive to male bees were long-chain alkanes and especially alkenes. These compounds, which are components of both the cuticular wax layer of the bee and the scent of the orchid, had not previously been regarded as important because of their relatively low volatility and almost ubiquitous occurrence in plants and insects.

Sexual mimicry is an inherently fascinating example of plant adaptation, and the topic has appealed to numerous researchers. New sexual mimicry systems continue to be discovered and are now known to occur outside the orchid family as

Floral Mimicry. Steven D. Johnson & Florian P. Schiestl.
© Steven D. Johnson & Florian P. Schiestl 2016. Published 2016 by Oxford University Press.
DOI 10.1093/acprof:oso/9780198732693.001.0001

Figure 5.1 Bertil Kullenberg, professor of entomology at the University of Uppsala, Sweden, who was a pioneer in the investigation of sexual mimicry. This photo shows him carrying out gas chromatographic analyses. Photo © Gunhild Kullenberg.

well (Ellis and Johnson 2010; Vereecken et al. 2012). Sexual mimicry is currently among the best understood pollination systems in terms of both the mechanisms of pollinator attraction and its consequences for evolutionary diversification. The molecular mechanisms of these phenomena have recently also come under scrutiny (see Chapter 8 for details). Research on sexual mimicry can be

approached from either a botanical or a zoological perspective, but the most successful studies have been those that combine the two.

In this chapter we review the occurrence of sexual mimicry among plants, and the most commonly recruited groups of pollinators. We show which signals are the most important, and how they evolve. We then review the evidence for pollinator shifts and how they may drive speciation in this highly specific pollination system. We also discuss how the typically low genetic differentiation between species of sexual mimics can be interpreted. Lastly, we consider the factors that drive the evolution of sexual mimicry in different plant lineages.

Phylogenetic distribution of sexual mimics and their pollinators

The overwhelming majority of examples of sexual mimicry have come from the orchid family. To date, 18 genera in the orchid subfamilies Orchidoideae and Epidendroideae have been conclusively identified as sexual mimics, and there is little doubt than many more will follow. Outside the orchid family, sexual mimicry has been demonstrated for only two

---

Box 5.1 Gas chromatography and antennal electrophysiology

Volatile blends emitted by flowers are usually very complex, and consist of both true signals and other compounds that are by-products of biosynthesis or have other functions. A possible reason for this may be that scent signals often evolve secondarily by modifications to pathways that produce existing volatiles with other functions. For example, cuticular waxes consist of hundreds of compounds (alkanes, alkenes, esters, aldehydes, and alcohols of different chain lengths) with the primary function of preventing desiccation. The signaling function of a subset of these compounds likely evolved secondarily to imitate the mating signals in solitary bees or recognition signals in social bees or ants. Therefore, volatiles may often represent a mixture of true signals and noise, making it necessary to screen complex volatile mixtures for those that may have a functional role. A method that has proved to be a powerful tool for this task is gas chromatography with electroantennographic detection (GC-EAD). This method allows for online monitoring of the responses of olfactory neurons to volatile organic compounds

(VOCs) eluting from a chromatographic column. The methodological setup consists of a gas chromatograph, a column splitter, a heated outlet, and a device for electroantennographic recording. During gas chromatography, a blend of compounds is separated into individual compounds with different retention times. This is achieved because compounds interact with the stationary phase inside the GC column with different intensities. During GC-EAD, the GC column is split at the end with one branch leading into a detector (flame ionization detector or mass selective detector) and one branch being directed over an insect antenna mounted between electrodes for electrophysiological recording. This method was originally developed for identifying insect pheromones (Moorhouse et al. 1969) but has been adapted for plant and floral volatiles (Schiestl and Marion-Poll 2002). Combined with behavioral assays this method has led to the identification of "behaviorally active" VOCs in several sexual mimicry systems (e.g., several *Ophrys* species, *Chiloglottis*, *Caladenia*, *Drakaea*).

plant species so far—*Gorteria diffusa* (Asteraceae) (Ellis and Johnson 2010) and *Iris paradoxa* (Iridaceae) (Vereecken et al. 2012). The reason why sexual mimicry is especially common among orchids probably lies in their morphological predisposition to it. Orchids have zygomorphic flowers, with one petal (the labellum) often being modified, not only as a landing platform for pollinators but also as a signal structure imitating a female insect's body. In addition, orchids with their highly efficient pollination systems that utilize pollinaria for pollen transfer are able to tolerate low rates of pollinator visitation, as is the case for many rewardless flowers, including those that are sexual mimics (see Chapter 2).

Plant sexual mimics occur in a wide range of habitats, from temperate grasslands to tropical rainforests, yet these mimics exploit just a small subset of the insect taxa available in these habitats. A surprisingly restricted set of Hymenoptera (bees, wasps, ants) and some Diptera (flies) serve as pollinators of these sexual mimics (Fig. 5.2) (Gaskett 2011). Lepidoptera (moths and butterflies)—one of the main orders that act as pollinators in the angiosperms—are not known to pollinate plant sexual mimics, even though some moths fall prey to spiders that chemically mimic their mating pheromones (Yeargan 1994). Within the Hymenoptera, the most commonly exploited groups are thynnine wasps

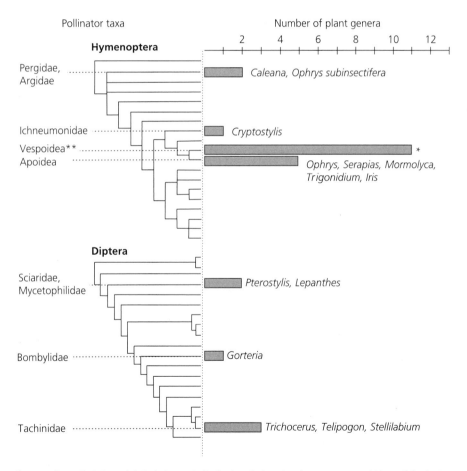

Figure 5.2 Pollinators of sexual mimics and their phylogenetic distribution. Phylogenies of Hymenoptera and Diptera following Yeates et al. (2003) and Vincent et al. (2010) are shown. Plant genera are given on the right (**Drakaea, Chiloglottis, Caladenia, Spiculaea, Arthrochilus, Paracaleana, Calochilus, Leoporella, Geoblasta, Disa, Ophrys*). Pollinator groups and their position on the respective phylogenies are indicated on the left (**Thynninae, Pompilidae, Sphecidae, Scoliidae, Formicidae). Pollinators of most plant genera are within the Vespoidea and Apoidea in the order Hymenoptera.

(Tiphiidae, Thynninae) and bees (Apidae), while among Diptera they are the bee flies (Bombyliidae) and fungus gnats (Sciaridae, Mycetophilidae), the latter being the pollinators of the species-rich orchid genera *Lepanthes* (Blanco and Barboza 2005) and *Pterostylis* (Phillips et al. 2014b). This pattern whereby sexual mimicry involves particular insect groups is reinforced by the independent recruitment of scoliid wasps (Hymenoptera, Scoliidae) as pollinators in the European *Ophrys speculum* (Pouyanne 1917), the Australian genus *Calochilus* (Fordham 1946), and the South American *Geoblasta* (Fig. 5.3) (Ciotek et al. 2006). Thus it seems that certain pollinator groups are predisposed to falling victim to sexual mimicry, and these groups are not necessarily typical pollinators of orchids in general. Thynnine and scoliid wasps, for example, are not commonly known as pollinators of other orchids (but see Stoutamire 1983). The reason for this pattern is largely unknown.

One possible explanation—namely constraints in the biochemical mimicry of the sexual pheromones of pollinators—seems increasingly unlikely as more of the enormous diversity of volatiles produced by orchids is discovered (Kaiser 1993). Some rewarding orchids are even known to produce compounds resembling moth sex pheromones (Jersáková et al. 2010), despite the curious absence of moths among pollinators of sexual mimics. The diversity of volatiles deployed by orchid sexual mimics is also impressive given the fact that these compounds have, as yet, only been identified in relatively few systems (Fig. 5.4). The morphology of

pollinators is unlikely to limit their exploitation by sexual mimics, as many Lepidoptera and Coleoptera are efficient pollinators of other orchids, both rewarding and deceptive.

Currently, the most likely explanation for the preferential use of certain pollinators by sexually deceptive plants is their mating behavior, particularly that of the males (Gaskett 2011). Male hymenoptera often patrol given routes in search of virgin females (Alcock et al. 1978), a behavior that when applied to sexual mimics appears to lead to reliable pollen transfer and long-distance pollen flow (Peakall and Beattie 1996; Peakall and Schiestl 2004; Whitehead and Peakall 2013). Long-distance pollen transfer is indeed one of the hypotheses for the evolution of sexual mimicry (see "The evolution of sexual mimicry"). Whether similar behavior is found in the fungus gnats and bombyliid flies involved in some sexual mimicry systems is unknown. Another aspect is the typical avoidance of sexual mimics after attempted copulation, which is likely to increase outcrossing and reduce self-pollination (Ayasse et al. 2000; de Jager and Ellis 2014). More work is needed to shed light on the highly skewed exploitation of pollinators by sexual mimics.

Signal evolution

The importance of olfactory signals

Volatile organic compounds are of key importance for sexual communication in insects. These sex pheromones are used mainly for detecting mates

Figure 5.3 Three flowers of sexual mimics that are pollinated by scoliid wasps. Long hairs on the margin of the labellum is a characteristic feature of all species. From left to right: *Bipinnula* (=*Geoblasta*) *pennicillata* (South America), *Calochilus campestris* (Australia) and *Ophrys speculum* (Europe). Photos © Santiago Benitez Vieyra, Florian Schiestl.

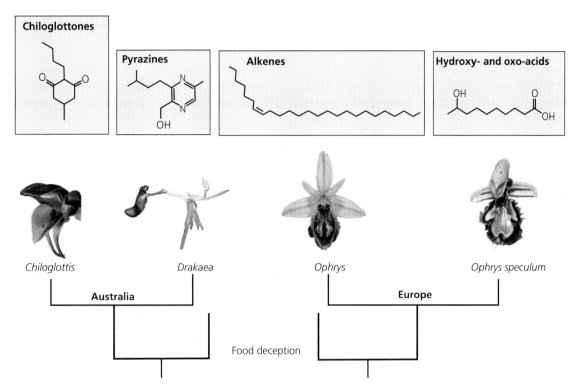

Figure 5.4 Volatiles attracting male pollinators produced by different groups of sexual mimics. Sexual mimics have independently evolved different chemical groups of compounds mimicking the pollinators' sex pheromone, indicating a lack of constraint on the production of those bioactive volatiles. Photos © Florian Schiestl.

over long distances and identifying receptive mating partners (Ayasse et al. 2001). Insects rarely rely on visual signals alone for mating because they have limited visual acuity and cannot use visual signals effectively for long-range communication and accurate species discrimination, particularly when species lack distinctive visual features. Chemical signals may often provide better acuity in intraspecific communication, and can be deployed regardless of the habitats or patterns of activity of the insects. Sex pheromones are often specific, enabling the identification of a conspecific individual within a species-rich insect community. Therefore it is not surprising that volatiles play a key role in sexual mimicry and often constitute the main attractive signal, both for long-distance attraction and for triggering copulation attempts with flowers.

Gas chromatographic analysis with electroantennographic detection (Box 5.1) has often been used to identify behaviorally active volatiles in insects (Moorhouse et al. 1969). This method can identify

volatiles that elicit responses in antennal olfactory neurons; these volatiles can later be tested for behavioral activity. In all systems where both female-produced sex pheromones and mimetic plant volatiles have been identified, the active compounds have proven to be chemically identical, though not always produced in the same amounts or relative proportions (Ayasse et al. 2003; Schiestl 2004). Orchid-produced "pseudopheromones" tend to be as specific as the actual pheromones, attracting only a few or even a single pollinator species. One possible exception to the general principles of chemical mimicry in sexually deceptive systems is the South African daisy *Gorteria*, where visual and tactile signals are sufficient to attract and induce landing (and sometimes also copulatory) behavior in male *Megapalpus* bombyliid flies (de Jager and Ellis 2012). Earlier studies showed that copulatory behavior can be induced in male *Usia* bombyliid flies in the Mediterranean using ink spots on pieces of paper placed in the field (Johnson and Dafni 1998), but this might occur only in the presence of the scent of conspecific female flies in that habitat.

The chemistry of sexual mimicry has proven to be highly diverse. One pattern is for genera of sexual mimics to show variation among species in the compounds used for pollinator attraction (Fig. 5.4). Another pattern is phylogenetic conservatism, where species within genera use very similar compounds. Closely related *Ophrys* species pollinated by *Andrena* and *Colletes* bees (Breitkopf et al. 2015), for example, tend to produce mixtures of alkanes and alkenes within the cuticular waxes (Ayasse et al. 2011). Species specificity with these relatively common compounds is achieved by the production of a specific pattern, i.e., relative proportions of the compounds in the blend. Of key importance are the abundances of alkenes with certain double-bond positions. For example, the *Andrena*-pollinated *Ophrys sphegodes* mostly produces alkenes with single double-bonds at the 9, 11, and 12 positions in the carbon chain, whereas the *Colletes*-pollinated *Ophrys exalata* primarily produces 7-alkenes (see Chapter 8 for further details). Although most—perhaps all—species of *Ophrys* produce alkanes and alkenes in abundance (Ayasse et al. 2011), some species have adopted other compounds as pollinator attractants. Two bumblebee-pollinated *Ophrys* species, for example, deploy polar, as yet unidentified compounds to deceive their pollinators (Gögler et al. 2011). Yet another species, the scoliid wasp-pollinated *Ophrys speculum*, produces oxo- and hydroxy-acids as sexual attractants (Ayasse et al. 2003). Even more diversity may be discovered within *Ophrys* in the future, as the chemical signals of many species, especially the large section of species pollinated by long-horn bees (e.g. the genera *Eucera* and *Tetralonia*), are still unknown.

The other chemically well-investigated group of sexually mimetic orchids belongs to the Australian Diurideae. Although most of these sexual mimics use thynnine wasps as pollinators, the chemical signals involved have proven to be more diverse than originally expected. The first system in which the pollinator-attracting chemistry was unraveled was the eastern Australian *Chiloglottis trapeziformis*. This orchid produces 2-ethyl-5-propylcyclohexan-1,3-dione, so-called chiloglottone, which is also the sex pheromone produced by the female wasp (see Figs. 5.4, 5.10) (Schiestl et al. 2003). This compound belongs to a formerly unknown class of natural products; different members of this "chiloglottone family," present in various combinations, were subsequently discovered in other species of *Chiloglottis* that attract thynnine wasp pollinators (Peakall et al. 2010). Although it seemed likely that chiloglottones are widespread sex pheromones among thynnine wasps, and that all thynnine-pollinated orchids would consequently produce them as attractants, this has proven not to be the case. Unexpectedly, some species of the Western Australian orchid genus *Drakaea*, also pollinated by thynnine wasps, use pyrazines as pollinator attractants (Fig. 5.4); these pyrazines were also shown to be sex pheromones of the pollinator wasps (Bohman et al. 2012, 2014). The fact that orchids have managed to successfully exploit both of these pheromone communication systems suggests that production of various chemical compounds, even highly unusual ones like chiloglottones, poses few (if any) constraints on them.

Visual signals

Many sexual mimics have dark, insectiform flowers or dark elevated spots, suggesting that visual signals play an important role in sexual deception (Figs 5.3 and 5.5). For humans, visual signals are of prime importance and we thus tend to overestimate their importance in plant–insect communication (Raguso 2008). Many older descriptions of sexual mimicry placed emphasis on the importance of visual signals. Since the work of Kullenberg, who

Figure 5.5 Mimic and model in an Australian sexual mimicry system: Left, a labellum of the orchid *Chiloglottis trapeziformis* and right, a female of the wasp *Neozeleboria cryptoides*. The dark knobs on the labellum, the so-called calli, are thought to mimic morphological properties of the female wasps. Photos © Florian Schiestl.

thoroughly demonstrated the primacy of scent over color in *Ophrys* pollination (Kullenberg 1961), the scientific focus in research on sexual mimics has been on olfactory signals (Schiestl 2005; Ayasse et al. 2011). Recently, however, the role of visual traits in sexual mimics has come under scrutiny again with the advent of approaches that incorporate the sensory abilities of pollinators. These investigations have focused on the functions of the colored display consisting of sepals and petals (excluding the labellum) in some species of *Ophrys*. These parts that make up the "perigone" (or "perianth") and are usually green, but sometimes white to pink (Spaethe et al. 2007). Intraspecific variation in perigone color is also found, such as in *Ophrys arachnitiformis*, where the perigone varies from green to white. In this species, however, the pollinator has been shown not to discriminate according to color, nor does scent co-vary with color, suggesting that the color variation evolved through relaxed selection (Fig. 5.6) (Vereecken and Schiestl 2009). This differs from species with a pink perigone being pollinated by long-horn bees. *Ophrys heldreichii*, for example, has a pink perigone. Although not directly involved in the mimicry of females (i.e., females have no pink coloration), this pink color increases

the attractiveness of the flowers to the bee pollinator (Fig. 5.7) (Spaethe et al. 2007) and is linked to higher pollination success in wild populations (Rakosy et al. 2012). But why does pink improve the attraction of flowers to male bees? Streinzer et al. (2009) suggested that the main function of a colored perigone is to increase green receptor contrast between the perigone and its background. Green receptor contrast is an important brightness channel for bees, making flowers more visible to bee pollinators. However, whether or not sexual mimics profit from increased green receptor contrast is likely to depend on the visual system of their pollinators. In some systems, the color of the perigone also plays a role in visual discrimination between different *Ophrys* species by pollinator bees (Streinzer et al. 2010). This was shown in *O. heldreichii* and *Ophrys dictynnae*, which differ in perigone color. In the absence of olfactory signals the pollinator of *O. heldreichii* is still attracted (but does not attempt copulation), and prefers intact *O. heldreichii* over *O. dictynnae* flowers. This preference disappeared when the perianth of both species was removed and replaced by pink-colored paper (Fig. 5.7). Because the two *Ophrys* species also differ in labellum pattern, and the pattern difference was still apparent in this experiment, the labellum pattern probably plays no role in visual discrimination of the species by pollinators. This is interesting because color patterns on labella are common among *Ophrys* and other sexual mimics and are often highly variable within species. The functions (if any) of this often striking variability in coloration are still largely unknown (Dickson and Petit 2006).

Visual signals in sexual mimics definitely merit more attention. For example, some *Ophrys* species, besides their pink perigone, have yellow margins on their labellum, which may similarly increase green receptor or color contrast. Alternatively, such color may have evolved through receiver bias when the pollinator bee is specialized on yellow flowers, thus having an innate preference for yellow (Spaethe et al. 2007; Vereecken and Schiestl 2009). Visual signals in mimics also evolve under selection for improving the similarity of model and mimic, such as in the Australian orchid genus *Cryptostylis* (Gaskett and Herberstein 2010). Other examples may include *Ophrys speculum* which sports orange hairs and a

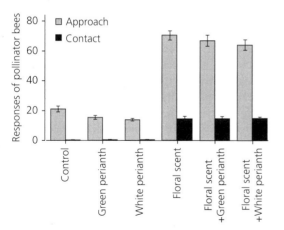

Figure 5.6 Responses of male *Colletes cunicularius* bees to scent samples and the floral perianth (flowers without labella, made odorless by washing with hexane) of *Ophrys arachnitiformis*. The graph shows that visual signals alone do not induce copulation attempts, nor do they enhance the attractiveness of scent. This differs from other *Ophrys* species, where visual signals play a more important role in pollinator attraction (see text). Adapted from Vereecken and Schiestl (2009).

(a)

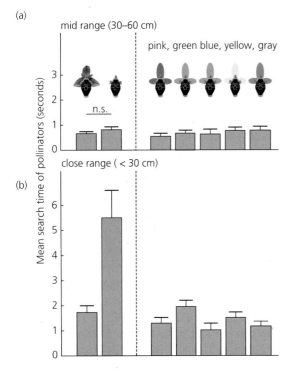

(b)

Figure 5.7 Search times of *Tetralonia berlandi* male bees approaching *Ophrys heldreichii* flowers at (a) mid-range (30–60 cm) and (b) close range (<30 cm). The left panels of the graph show a comparison of real flowers with those in which the perianth was removed. The right panels show flowers in which the original perianth was replaced with artificial perianths of different colors. In the mid-range, the perianth does not influence search time; at close range, lack of a perianth greatly increases search time, but the addition of an artificial perianth again improves search time. (n.s.: not significant). Adapted from Streinzer et al. (2009).

blue shiny center (Plate 3a), thereby closely matching the appearance of females of its pollinator, the scoliid wasp *Campsoscolia ciliata* (Plate 3b). Similar colors are found in *Calochilus*, a genus of Australian diurid orchids that is also pollinated by scoliid wasps. In some species of *Ophrys*, variation in the usually dark insectiform labellum ranging from pitch black to light brown matches the coloration of the body of the pollinator bee, at least to the human eye (Schlüter and Schiestl 2008). In several systems, however, overall visual similarity between flowers and their female insect models is quite low. A very loose match, for example, is found between overall floral shape in orchid genera such as *Caladenia*, *Cryptostylis*, or *Chiloglottis* and their corresponding

female wasp models (Fig. 5.5). Species that have independently evolved to exploit identical pollinator species sometimes produce flowers that are very different in shape, such as *Caladenia* and *Drakaea* (Phillips et al. 2013), and different species of *Cryptostylis* that share a single pollinator species (Gaskett 2012). It is thus likely that some traits such as overall shape play little role in deceiving pollinators in these systems.

Nevertheless, some interesting studies have focused on the role of morphological traits such as shape, size, and height of flowers. In the Australian diurid *Chiloglottis trapeziformis*, size does seem to matter for its thynnine-wasp pollinators (Schiestl 2004): A dummy larger than a female elicited more copulation attempts than a smaller dummy approximately the same size as a female, suggesting that sexual mimics can increase their attractiveness by increasing the size of the floral part that imitates a female insect. In a selection study, Benitez-Vieyra et al. (2009) showed that pollinators do not necessarily select for closer resemblance of model and mimic in terms of dimensions. Rather, size and novel shape seemed to be favored, albeit to varying degrees in different years. A study by de Jager and Peakall (2016) showed that size and shape differences mattered for copulation attempts by different thynnine pollinators attracted to *Chiloglottis valida* and *C. trapeziformis*. This study also highlighted the importance of selection for mechanical fit as well as for mimicry. These two orchids differ in peduncle length, thus presenting their flowers at different heights. Thynnine pollinators are known to discriminate among heights of *Chiloglottis* flowers (Peakall and Handel 1993), and in this particular species pair one pollinator prefers flowers at a lower height whereas the other prefers flowers at an elevated height, thus likely selecting for differences in peduncle lengths (Schiestl and Peakall 2005).

Tactile stimuli

Insect-mimicking parts of a flower can play different roles in the pollination of sexual mimics (Fig. 5.3). In particular, such structures seem to improve the trait matching between model and mimic (Kullenberg 1961), as exemplified by the hairs on the labellum margin in the scoliid wasp-pollinated

Box 5.2 Bioassays

In mimicry research, bioassays with operators are often a critical component of a successful investigation. In floral mimicry systems, bioassays can be used to quantify the attractiveness of a whole flower or specific parts thereof, to test the degree to which pollinators discriminate between model and mimic, and to quantify search time, visitation time, pollen export, etc. Bioassays can be done in the field or with captive pollinators in a cage or wind tunnel. The advantage of using pollinators in the field is the natural conditions under which the animals are tested, but the downside is the impossibility of controlling for factors like their experience and age. Cage experiments with captive-reared pollinators offer great opportunities to control environmental factors and to compare naïve and experienced pollinators, but suffer from a somewhat artificial setting, including the lack of a natural flower community. The setup of a cage experiment can be critical. For example, Forrest and Thomson (2009) showed that background complexity greatly affects the behavior of pollinators in a cage.

One type of bioassay involves testing a pollinator's response to a stimulus by counting pollinator approaches to real or dummy flowers in arrays. The effectiveness of the stimulus should ideally be compared with both positive and negative controls. The positive control is usually a real flower and serves to assess the presence, activity, and natural behavior of pollinators. The negative control lacks the stimulus of interest but is otherwise identical to the dummy flower presenting this stimulus. Another type of bioassay is designed to test pollinator choices in dual-choice assays, such as presentation of paired inflorescences using a "bee interview stick" in the field (Fig. 1) or a Y-maze olfactometer test. Depending on the research questions addressed, natural or artificial flowers can be used. Natural flowers will be used, for example, to test whether pollinators initially discriminate between model and mimic in order to verify a mimicry hypothesis (Johnson 2000). Flowers can also be manipulated to estimate the relative importance of traits, for example through removing flower parts and subsequently quantifying attractiveness (Streinzer et al. 2009). Johnson et al. (2003) manipulated the inflorescence of a floral mimic and showed that the spatial arrangement of flowers plays an important role for deceiving operators in this mimicry system. Floral signals can also be added, for example Schiestl and Ayasse (2001) dripped small amounts of farnesyl hexanoate diluted in solvent onto intact *Ophrys* flowers and offered them to pollinators to test the behavioral effect of the volatile. In general, artificial flowers offer

Figure 1 The pollinator interview technique, whereby insects are offered a binary choice between flowers mounted on a T-shaped stick, allows insect discrimination between models and mimics to be studied under field conditions. Photo © Steven Johnson.

excellent opportunities for manipulating individual traits, and have the advantage that they do not vary in traits other than those of interest, but they often only crudely resemble their real versions and thus may suffer reduced attractiveness. However, Jersáková et al. (2012) developed artificial flowers that, given the right color, matched real flowers in terms of attractiveness to fly pollinators (see Fig. 4.4). An interesting new method for constructing realistic artificial flowers for bioassays is to use 3D printing with silicone (Policha et al. 2016). This method will allow much more detailed assessment of the function of morphological traits and manipulation of the color and scent of artificial flowers that have a realistic depiction of the floral morphology.

When testing floral color, it is important to assess color properties with spectrophotometry and even by using insect vision models (see Box 3.1). Besides spectral reflectance, one also needs to control for translucence, brightness, and contrast with the background as these are important factors for pollinator vision (Chittka and Kevan 2005).

continued

Box 5.2 *Continued*

The challenge with adding scent to flowers is to obtain a continuous rate of evaporation of scent in the right concentration. Continuous delivery of a fixed amount can be achieved by dispensing scent volatiles dissolved in solvent through a wick placed inside a narrow glass capillary, but this has the disadvantage that solvent is always present and may interfere with behavior. In some cases a solvent mixture can be pipetted directly onto a solid surface (Fig. 2). This can work if the active compounds are detected even in trace amounts, such as chiloglottones, or have low volatility, as is the case for alkanes and alkenes in *Ophrys*. After application, solvent will evaporate within a few minutes while the heavier compounds will volatilize much more slowly. More volatile compounds will not persist much longer than the solvent, making frequent recharging necessary and testing times short. Various methods have been used to achieve a continuous evaporation, for example dissolving volatiles in mineral oil (a blend of higher alkanes). Mineral oil or other long-lasting solvents, will, however, be present together with the scent compounds of interest, and even if a mineral oil control shows no behavioral activity on its own, one always tests the interaction between a test compound and the solvent it is diluted in, rather than the activity of the pure compound. Numerous studies have shown that alkanes can be smelled by insects and can influence the activity of other volatiles (Schiestl et al. 2000). A method that allows continuous evaporation over a longer time without solvent is the use of rubber septa; after being soaked in a scent solution and dried for some time these continue to emit the scent compound over several hours (Waelti et al. 2008). While this method is good for adding scent compounds to real or artificial flowers, the removal of scent compounds is considerably

more difficult and will typically require genetic manipulation through silencing of odor genes (Kessler et al. 2015), a technique not yet established for any mimicry system.

Figure 2 Field bioassays can be performed with small amounts of the compound chiloglottone (the active compound emitted by the Australian sexual mimic *Chiloglottis trapeziformis*) applied directly onto black plastic beads. The pollinator, *Neozeleboria cryptoides*, is readily attracted to the smell and alights on the dummy. In this study, the preferences for size and intensity of scent were examined by using beads of different sizes with varying amounts of chiloglottone (Schiestl 2004). Photo © Florian Schiestl.

mimics *Ophrys speculum*, *Calochilus*, and *Geoblasta* that resemble the hairs on the body of female wasps (Plate 3a). In the South African daisy *Gorteria*, raised insect-like ornaments on the florets (Plate 4h) play a key role in pollinator attraction, probably through both tactile and visual stimulation (de Jager and Ellis 2012). In *Ophrys*, the direction of the labellar trichomes—either facing from the stigma to the labellar tip or vice versa—influences whether the male bee pseudocopulating on the labellum adopts a position with its face toward the stigma or toward the labellum tip (Ågren

et al. 1984; Paulus and Gack 1990). Thus, the labellar trichomes may have some influence on morphological isolation between *Ophrys* taxa, a principle that may apply more generally within sexual mimics (Phillips et al. 2014b).

Receiver bias

Signal evolution in sexual mimics, as for mimics in general, is driven by preferences and sensory abilities in the operator (dupe), which in this case is a pollinator. In protective mimicry in animals, as well

Plate 1 Examples of generalized food deception. (a) Close-up of the yellow and purple forms of the European food-deceptive orchid *Dactylorhiza sambucina*. Photo © Herbert Stärker. (b) Color polymorphism in a population of *D. sambucina*. Photo © Herbert Stärker. (c) A queen bumblebee (*Bombus lapidarius*) probing the empty floral spurs of the European orchid *Anacamptis morio*. Photo © Steven Johnson. (d) Pollen-imitating hairs on a sepal of the Mediterranean *Iris germanica*. Photo © Florian Schiestl. (e) Pollen-like nectar guides on the labellum of the African orchid *Eulophia cucullata*. Photo © Herbert Stärker.

Plate 2 Examples of Batesian food-source mimicry. (a) The non-rewarding South African orchid *Disa pulchra* (left) is a mimic of the rewarding iris *Watsonia lepida* (right). Long-tongued flies do not discriminate between the two species when foraging for nectar. (b) A related orchid, *Disa ferruginea*, is also non-rewarding and relies on mimicry to attract its pollinator, the mountain pride butterfly, *Aeropetes tulbaghia*. (c) The iris *Tritoniopsis triticea* is the model for the red-flowered form of *D. ferruginea*. Arrows in (b) and (c) indicate pollinaria of *D. ferruginea* on the proboscis of the butterflies. Photos © Steven Johnson.

Plate 3 Examples of sexual mimicry in European orchids. (a) *Ophrys speculum* (= *O. ciliata*) and (b) a female scoliid wasp *Campsoscolia ciliata* (= *Dasyscolia ciliata*). The females are the model in the sexual mimicry employed by *O. speculum*. (c) A male *C. ciliata* attempting copulation on a flower of *O. speculum*. (d) A male *Andrena pillipes* bee attempting to copulate with a flower of *Ophrys garganica*. Pollinaria of the orchid are attached to its head. (e) A male *Anthophora atroalba* bee attempting to copulate with a flower of *Ophrys omegaifera*. The position of the bee differs from that adopted by bees on the other *Ophrys* species shown here; as a consequence pollinaria of *O. omegaifera* are attached to the bee's abdomen. All photos © Nicolas Vereecken, except (d) © Florian Schiestl.

Plate 4 Examples of sexual mimicry. Closely related members of the *Ophrys insectifera* group, (a) *O. insectifera*, (c) *Ophrys aymoninii*, and (e) *Ophrys subinsectifera*, have adapted to three different pollinators, namely (b) the digger wasp *Argogorytes mystaceus*, (d) the solitary bee *Andrena combinata*, and (f) the sawfly *Sterictiphora gastrica*, respectively. (g) *Disa atricapilla*, a sexually deceptive South African orchid which is pollinated by male sphecid wasps. (h) The South African daisy *Gorteria diffusa* attracts male bombyllid flies through mimicry of female flies resting on the florets. Photos (a)–(f) © Nicolas Vereecken, (g), (h) © Steven Johnson.

Plate 5 Examples of sexual mimicry in Australian orchids. (a) Flowering plants of *Chiloglottis* aff. *valida*. (b) A flower of *Chiloglottis trapeziformis* and (c) its pollinator, a male thynnine wasp, *Neozeleboria cryptoides*, attempting copulation. (d) *Chiloglottis seminuda* with its thynnine pollinator *Neozeleboria* sp. (e) *Chiloglottis* aff. *jeanesii* with its pollinator *Neozeleboria* aff. *impatiens*. All *Chiloglottis* species illustrated here use different combinations of chiloglottone volatiles to attract their pollinators. (f) A male thynnine wasp, *Zaspilothynnus nigripes*, attempting to carry away the hinged female decoy of a flower of *Drakaea livida*, (g) making contact with the column, and (h) flying away with freshly affixed pollinaria. All photos © Rod Peakall.

Plate 6 Examples of oviposition-site mimicry. (a) Flowers of the African parasitic plant *Hydnora visseri* emerging above ground. This species is a carrion mimic that attracts necrophagous beetles and imprisons them in a below-ground chamber. Photo © Jay Bolin. (b) The hairy spathe and appendix of the Mediterranean arum *Helicodiceros muscivorus*, which is pollinated by carrion flies. (c) Thermal image showing heating of the appendix. (d) Opened floral chamber of *H. muscivorus*. (e) Thermal image showing heating of the male florets. Photos (b)–(e) © Kikukatsi Ito. (See Figure 6.2 on page 99).

Plate 7 Examples of oviposition-site mimicry. (a) The Asian parasitic plant *Rafflesia kerrii*, a carrion mimic belonging to a plant lineage that contains the world's largest flowers. Photo © Hans Bänziger (b) The African succulent and carrion mimic *Stapelia grandiflora*. Photo © Steven Johnson. (c) A blowfly and freshly laid eggs in the flower of *Stapelia gigantea*. Photo © Steven Johnson. (d) *Jaborosa rotacea* (Solanaceae), a carrion mimic native to southern South America. Photo © Andrea Cocucci. (e) The Asian orchid *Bulbophyllum subumbellatum* is pollinated by flesh flies. Photo © Ong Poh Teck. (f) Dozens of blow flies attracted to a flowering plant of *Bulbophyllum lasianthum* in Malaysia. Photo © Ong Poh Teck. (g) The stinkhorn fungus *Clathrus archeri* attracts flies through a combination of carrion and fecal odors. Photo © Steven Johnson. (See Figure 6.9 on page 108).

Plate 8 Examples of oviposition-site mimicry. (a) The North America pipevine *Asarum caudatum* is pollinated by fungus gnats. Photo © Robert Carr. (b) The Neotropical orchid *Dracula lafleurii*, a fungus mimic that attracts drosophilid flies. Photo © Barbara Roy. (c) The Asian orchid *Paphiopedilum callosum* mimics aphids and attracts aphidophagous hoverflies. Photo © Steven Johnson. (d) Aphid mimicry also occurs in *Paphiopedilum rothschildianum*. The arrow indicates the position of one of many hoverfly eggs laid on the staminode. Photo © Rogier van Vugt. (e) *Phragmipedium pearcei*, a South American slipper orchid pollinated by aphidophagous hoverflies, one of which can be seen emerging from the flower chamber. Photo © Alex Portilla. (f) The Eurasian orchid *Epipactis veratrifolia* uses the volatile signature of aphid alarm pheromones to attract aphidophagous hoverflies. Photo © Johannes Stökl.

as food-source mimicry in plants, selection for increased similarity between model and mimic can be expected because such similarity reduces the likelihood that operators will be able to discriminate between the mimic and the model on which it gained experience. The situation may be somewhat different in sexual mimicry, where mimics exploit mostly innate sexual responses of a single (or at most a few) operator species. Thus the sensory system of the operator, including higher-level neuronal processing that mediates the behavior, will shape selection on floral signals.

Several studies in sexual mimics have shown that this selection does not necessarily favor a perfect match in signals between mimic and model. Indeed, pollinators may actively favor dissimilarity between sexual mimics and their models. In the South American orchid *Geoblasta pennicillata*, for example, selection on floral shape was found to favor uncommon phenotypes above those with a perfect match to female wasps (Benitez-Vieyra et al. 2009). In a study focusing on the role of scent in attracting bee pollinators of *Ophrys*, Vereecken et al. (2007) showed that male pollinators prefer odor bouquets that differ from those of local female bees. Local orchids also differ in scent from local female bees (their models), and this makes them more attractive to male pollinators than the co-occurring females (Fig. 5.8) (Vereecken and Schiestl 2008). Increased attractiveness of the orchids' perfume was mainly due to the increased production of (Z)-7-alkenes in the flowers, the key compounds in the bees' sex pheromone. Higher attractiveness of orchids compared with females was also shown in two other systems, in both of which the orchids produce higher amounts of pollinator-attracting compounds (Ayasse et al. 2003; Schiestl 2004). Collectively, these are examples of how receiver bias in pollinators, preferring different shape, scent bouquets, or higher amounts of scent, can drive the evolution of floral signals in sexual mimics. The result is sometimes imperfect mimicry characterized by a level of dissimilarity in signals between model and mimic.

The above examples show that plants can evolve "supernormal signals" in response to pollinator-mediated selection. But why can plant mimics become more attractive than their respective female models? Two aspects are relevant here. Firstly,

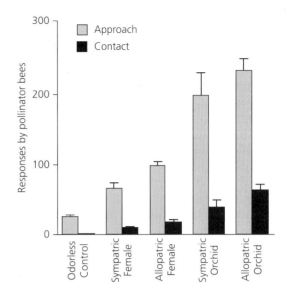

Figure 5.8 Attractiveness of different scent samples to male *Colletes cunicularius* bees, the pollinator of several species of *Ophrys* in the Mediterranean. The graph shows that the scent of the orchids is more attractive than the scent of female bees. In addition, allopatric (from different populations) females and orchids tended to be more attractive than sympatric ones (from the same population). Adapted from Vereecken and Schiestl (2008). Copyright (2008) National Academy of Sciences, USA.

plants do not face the same trade-offs that insects usually have to contend with in pheromone communication. One such trade-off for insects is the possibility of parasites "eavesdropping" on pheromone signals (Fatouros et al. 2005). Secondly, selection on attractiveness in female insects is likely to be weaker than in orchid flowers, because males are expected to compete vigorously for access to females, and females are probably not limited in their reproductive success by attracting males. In orchids, however, this is quite different, as pollination success is typically low and is limited by the attraction of pollinators. Thus higher attractiveness is expected to increase pollination success, and orchids should evolve a maximally attractive display, as long as high attractiveness will not compromise fitness through self-pollination.

An often-asked question is whether it is possible for male pollinators to evolve counter-adaptations against orchids, for example through an improved ability to discriminate between females and mimics. The commonly observed avoidance behavior shown by many pollinators of sexual mimics after pseudocopulation (Ayasse et al. 2000) is not likely to reflect a counter-adaptation that evolved in response to sexual mimics, however, because this is a normal part of the mating behavior found in many insects (Ayasse et al. 2001). Any potential adaptations in pollinators for avoiding orchids would involve a process of antagonistic coevolution (an "arms race" scenario) similar to the expected evolutionary dynamics in host–parasite interactions. Whether these counter-adaptations are favored would depend on the fitness costs that mimics impose on the male pollinators. Such costs could be time wasted on flowers or increased mortality due to pollinia load or predation on flowers. These factors may often be negligible; certainly visits to orchids do not prevent mating success in males, as marked males visiting flowers have been observed mating with females (Peakall 1990). Some studies have shown that males could indeed miss mating opportunities while visiting mimics or when avoiding areas where mimics grow (Fig. 5.9) (Wong and Schiestl 2002; de Jager and Ellis 2014), or waste sperm when ejaculating on flowers (Gaskett et al. 2008). Obviously, selection for counter-adaptations in male pollinators cannot target the pheromone signal itself because individuals ignoring sex pheromones would not find mates.

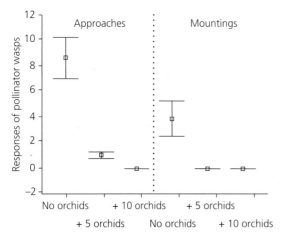

Figure 5.9 Responses of male *Neozeleboria cryptoides* wasps to their conspecific, calling females without and with orchids added in an experimental arena. The figure shows that with orchids present, males respond less to their conspecific females, probably because they avoid spots where they have attempted to copulate with flowers. From Wong and Schiestl (2002).

Avoidance would thus have to involve plant-specific cues such as labellum shape or non-pheromonal volatiles. Counter-adaptations would also come at a cost for males, as individuals that are very choosy may again miss mating opportunities because of speed–accuracy trade-offs (Chittka et al. 2009), and choosy individuals may also refuse to mate with females that deviate too much from the population mean in terms of their mating signals. Thus the evolution of counter-adaptations in pollinators of sexual mimics may only occur if the costs imposed by the mimics are severe, thus fuelling selection to actively avoid mimics. We think that counter-adaptations of pollinators to avoid sexually deceptive orchids are unlikely, but more studies that quantify costs imposed by mimics and identify possible counter-adaptations in pollinators are needed. For example, the spatial and temporal overlap between mimics and models need to be better assessed to quantify possible negative effects of mimics on models.

Pollinator shifts and speciation in sexual mimics

Species diversification among sexual mimics is astonishing, and some groups have undergone recent adaptive radiations. The genus *Ophrys*, for example, has diversified in the last few million years, driven by shifts between pollinator groups (Breitkopf et al. 2015). Sexual mimicry is among the most specialized of all pollination systems, with many sexual mimics having only one known pollinator species. This specificity spawned early ideas about pollinators mediating reproductive isolation and driving speciation (Paulus and Gack 1990). The idea that different pollinators, or differences in pollen placement on shared pollinators, can act as pre-pollination reproductive barriers (floral isolation) among plant species goes back to a seminal paper by Verne Grant (1949) and was later incorporated into the Grant–Stebbins model of pollinator-driven speciation (Grant and Grant 1965; Stebbins 1970).

Floral isolation is usually divided into ethological isolation—the attraction of different pollinator species by flowers—and morphological isolation—the differential placement of pollen on the body of the same pollinator species (Grant 1994). Both types of floral isolation are important in sexual mimicry.

Several studies have shown that floral isolation is strong in sexual mimics, and sometimes even represents the only reproductive barrier among species (Xu et al. 2011; Peakall and Whitehead 2014; Sedeek et al. 2014). Species definitions are often problematic in sexual mimics, and morphological or genetic differences are not always apparent between seemingly isolated plant lineages.

The species problem

The definition of a species has been an ongoing challenge in the study of sexual mimicry. Conventional taxonomic practice is to define and describe species by using traits that differ consistently between groups of organisms (i.e., using traits that can be used to "diagnose" a species reliably), and these traits typically co-vary with genes responsible for species-specific adaptations and/or reproductive isolation. The basic premise is that new lineages, that is, those that are reproductively isolated, will develop distinct traits, and that some of these traits will in turn promote reproductive isolation. The challenges in the case of sexual mimics are, firstly, that their high variability in floral morphology (*Ophrys* being a good example) makes it difficult to draw neat conventional species boundaries and, secondly, that different species can have similar morphology or even DNA sequences used for barcoding, yet be reproductively isolated because of differences in floral scent. This situation is a likely consequence of floral scent being the key trait for maintaining reproductive isolation in sexual mimics. Floral scent, however, is a problematic trait for use in taxonomy because it is identifiable only by chemical analysis in the laboratory and most taxonomists accept only those traits that are measurable by "ordinary means" (Cronquist 1988).

As a consequence, the taxonomy of sexual mimics has remained problematic and controversial. An orthodox application of the morphological species concept can lead to excessive splitting due to the high variability found in some taxa (Deforge 1995). The description of too many species is undesirable because it leads to confusion, making subsequent identification very challenging. At the other extreme, genomic or phylogenetic approaches can lead to excessive lumping of species due to their similarity on

a whole-genome basis (Devey et al. 2009). Even molecular approaches can fail to detect evolutionarily independent units, as closely related species often show very little genomic differentiation yet can be strongly reproductively isolated and differ in a few genes under selection (Sedeek et al. 2014; Whitehead and Peakall 2014). This problem of lack of molecular differentiation among likely species is not confined to sexual mimics and is well known from other plant groups as well, such as iceplants and larkspurs (Hodges and Derieg 2009).

One approach to the problem, based on a strict interpretation of Ernst Mayr's biological species concept, has been to use pollinators to define species boundaries in sexual mimics (Paulus and Gack 1990; Hopper and Brown 2007). The reasoning is that pollinators of these sexual mimics maintain reproductive isolation and thus are relevant for delimitation of species. It could also be argued that pollinators are responding to real phenotypic differences among taxa in terms of volatile emissions. While this may be useful for distinguishing similar, sympatric species, a general problem is that pollinator attraction may not be consistent on a geographic scale due to local differences in the preferences of pollinators for mating signals (Vereecken et al. 2007). It is also possible for reproductive isolation to break down when flowers of sexual mimics are visited indiscriminately by other insects such as beetles (Steiner et al. 1994). Therefore none of these approaches can be applied satisfactorily, and a simple solution to these taxonomic problems has remained elusive. We advocate a multidisciplinary approach that involves detailed investigations of morphology, floral scent, pollinator attraction, and population genetics (and genomics where feasible). Up to now, few studies have gone to this degree of effort, but some have nicely shown that that this approach can lead to the identification of morphologically cryptic taxa in the early stage of divergence (Mant et al. 2005b; Peakall et al. 2010; Peakall and Whitehead 2014; Sedeek et al. 2014; Smouse et al. 2015).

Why are species of sexual mimics so similar genetically?

It is now well established that besides morphological similarity among species, genetic differentiation across the genome is typically low among sexually deceptive orchid species. Population genetic studies usually reveal high degrees of heterozygosity, high numbers of alleles in neutral molecular markers, and low differentiation between species (low F_{ST} values), indicating the sharing of polymorphisms among species. As a consequence, molecular phylogenies at the species level have proved difficult to establish (Soliva et al. 2001; Bateman et al. 2003; Devey et al. 2008; Peakall et al. 2010) and the taxonomy is controversial, as discussed above. Yet groups of plants (probably belonging to one species) that are specialized for a particular pollinator taxon can often be found (Paulus and Gack 1990; Bower 1996), and pollinator-mediated floral isolation between these plants and other groups of related species is strong (Xu et al. 2011; Whitehead and Peakall 2014). These patterns, common to European and Australian sexually deceptive orchids, have attracted two controversial interpretations, namely: (1) frequent gene flow between species and (2) recent divergence with differentiation at few loci (genic speciation) but little genomic differentiation as a consequence of shared ancestral polymorphisms. These two interpretations make different assumptions about evolutionary processes such as mechanisms of speciation and strengths of reproductive isolation, and have different implications for species descriptions. Yet the two evolutionary scenarios are difficult to discriminate because they leave similar traces in the plant genome.

Studies advocating gene flow between species as a result of hybridization have mostly dealt with the European genus *Ophrys*, with its often bewildering variability among individuals. In fact, the question of whether this high variability in *Ophrys* is due to species variation or hybridization dates back to the nineteenth century (Moggridge 1869), and has plagued field taxonomists ever since. Indeed, hybridization has been documented among *Ophrys* species (Stökl et al. 2008; Cortis et al. 2009), but the question here is whether hybridization is a local phenomenon occurring at similar levels to those in other plant groups, or whether gene flow is common between established species in the genus. One of the first studies to apply molecular markers to the question of hybridization in *Ophrys* was conducted by Soliva and Widmer (2003). They found

highly variable markers but low differentiation between species in the *O. sphegodes* group. Several taxa clustered according to their sympatric occurrence rather than taxonomic affinity, leading the authors to suggest that gene flow was common across species boundaries. In a combined phylogenetic/population genetic approach, Devey et al. (2008) used internal transcribed spacer (ITS), plastid DNA, and amplified fragment length polymorphism (AFLP) markers to define species boundaries. Similarly to Soliva and Widmer (2003), they inferred gene flow between existing species from the finding of a lack of genetic differentiation. Later, Devey et al. (2009) reached similar conclusions by studying 10 species of the *O. fuciflora* group from different locations in England and Europe using AFLP markers. Again, little genetic differentiation was found, and by advocating a "genetic species concept" the authors made the radical taxonomic suggestion of reducing the group to only two hypervariable species or subspecies.

Studies that supported a recent divergence scenario were conducted in the European genus *Ophrys* as well as the Australian *Chiloglottis*, the latter being characterized by little phenotypic variation but the occurrence of morphologically cryptic species. An approach using a combination of molecular markers, floral scent, and pollinator behavior was pioneered for *Ophrys* by Mant et al. (2005b) who, like Soliva and Widmer (2003), also found a lack of genetic differentiation in species from the *O. sphegodes* group. The analysis of floral odor, however, showed clear separation of species, together with specific pollinator attraction. These findings led the authors to conclude that a lack of genetic differentiation should not be taken automatically to indicate the existence of ongoing extensive gene flow.

Similarly, in another study using AFLP markers in the Australian genus *Chiloglottis*, Mant et al. (2005a) found little taxon-specific differentiation in sympatric species. Pollinator attraction, however, proved to be highly specific, and a detailed population genetic analysis showed sympatric taxa to be largely distinct. On the basis of these findings, sympatric divergence of species was suggested. Recent divergence is also supported by recent studies directly focusing on reproductive isolation in *Ophrys* and *Chiloglottis* (Xu et al. 2011; Sedeek et al. 2014;

Whitehead and Peakall 2014). In both genera, floral isolation mediated by specific pollinator attraction was found to be very strong, and was the primary barrier to gene flow among species. In *Ophrys* this was demonstrated in several subsequent seasons by monitoring the transfer of stained pollinia in combination with artificial crossings (Xu et al. 2011; Sedeek et al. 2014). In the study on *Chiloglottis*, pollinator attraction to flowers and bioassays with synthetic volatiles were employed (Whitehead and Peakall 2014). This study detected strong floral isolation, and a subsequent detailed investigation of genetic differentiation also found no evidence of gene flow across species boundaries (Whitehead et al. 2015). In their study of the Australian genus *Drakaea*, Menz et al. (2015) found that two ecotypes attracted different pollinators. The genetic differentiation between them was low but could be detected with Bayesian structure analysis, suggesting this to be a case of incipient speciation. Perhaps the most detailed and convincing analyses of species differentiation were recently presented by Peakall and Whitehead (2014) and Smouse et al. (2015). They applied new hypervariable chloroplast and nuclear DNA markers to four *Chiloglottis* species. Their population genetics approach was combined with analyses of floral odor and morphology. Consistent with earlier studies, they found low to moderately high genetic differentiation. They approached the problem of gene flow versus recent divergence using a haplotype-sharing analysis between two genetically very similar taxa, and found that haplotype sharing was considerably lower (and actually absent at the population level) than predicted under the null model of lack of reproductive isolation. Thus pollinator-mediated reproductive isolation is strong in these taxa, and chemistry tends to define species boundaries. Peakall and Whitehead (2014) also suggested that the low genetic differentiation between species is a consequence of large effective population sizes in combination with clonality, making genetic drift ineffective in causing neutral differentiation. Interestingly, orchids have generally been shown to exhibit low population genetic differentiation, perhaps due to long-distance seed dispersal combined with longevity and clonal reproduction (Phillips et al. 2012). Lastly, Sedeek et al. (2014), using a genotyping by sequencing approach,

showed that despite a broad overlap in most of the genome there is strong species differentiation in a few loci in species belonging to the *O. sphegodes* group. This finding, together with previous evidence for strong floral isolation and differences in floral odor, led the authors to suggest a genic speciation scenario at the early stage of the speciation continuum.

In summary, we feel that the available evidence supports the recent divergence scenario better than the hybridization scenario. In the Australian *Chiloglottis*, in particular, there is now overwhelming evidence in favor of the recent divergence scenario. Indeed, hybridization has only rarely been documented in this genus (Peakall et al. 1997). Hybridization does occur in the European *Ophrys*, but genic differentiation and strong floral isolation in sympatric species have also been shown. The basis for the commonly found high variability in floral morphology in this genus is still unknown. Schiestl (2005) has suggested a "converge and diverge" scenario (Grant et al. 2004), with temporal and/or spatial dynamics in reproductive isolation, depending on fluctuations in the pollinator environment. This idea merits further attention.

Speciation

Views about the process of speciation obviously depend on how species (the end-points of speciation) are defined, and this has been particularly controversial for sexual mimics. Most biologists consider speciation to be the process leading to the establishment of reproductive isolation among groups of individuals, with the consequence of (more or less) isolated gene pools. Others argue that species must also show phenotypic differences from other species, but there is always uncertainty about whether the phenotypic differences between species arise before, at the same time as, or after the development of reproductive isolation. Here we focus on the development of reproductive isolation through pollinator shifts in sexual mimics and then consider its link with traits such as scent chemistry that mediate these shifts.

Many species of sexual mimics are characterized by genetic similarity and strong pre-pollination reproductive barriers (floral isolation), mediated by specific pollinator attraction (Schiestl and Schlüter 2009) combined with weaker post-pollination barriers (Scopece et al. 2007). A rare exception to this rule is the Australian genus *Cryptostylis*, where related species share one pollinator insect and are genetically incompatible (Stoutamire 1975). Overall, the evidence implicates pollinator shifts as a key driver of speciation in sexually deceptive orchids. Obviously other forms of speciation, such as habitat adaptation or polyploidization (Amich et al. 2007), may also exist, but our focus here is on pollinator shifts because they arise from modifications to the floral phenotype (traits related to floral mimicry) and have an obvious link to the initiation of reproductive isolation. In sexual mimics, pollinator shifts can result in strong floral isolation due to the high pollinator specificity of these plants (Xu et al. 2011; Whitehead and Peakall 2014). This specificity is inherent to sexual mimicry because it is based on the specificity typically found in the responses of insects toward their mating signals.

A recent phylogenetic approach to patterns of diversification in *Ophrys* suggested that the rapid radiation of this genus in the Pleistocene was fuelled by shifts between different groups of bee pollinators (Breitkopf et al. 2015). A plausible driver for these pollinator shifts is the generally low and spatially variable pollination success typical in sexual mimics (Scopece et al. 2010). If a mutation resulted in the attraction of a different and more locally abundant insect pollinator species, and thus led to an increase in pollination rate and fecundity, then it would be likely to spread toward fixation in the population. Interestingly, pollination success sometimes correlates negatively with population size/density of plants (Johnson et al. 2012), particularly in rewardless plants (Fritz and Nilsson 1994), suggesting that plants in these large populations may show greater competition for pollinator attraction (Phillips et al. 2014a). Under these circumstances, mutations that attract new pollinators may be favored through a negative density-dependent process of selection (Waser and Campbell 2004). Low pollination success can also be the consequence of generally unfavorable plant–pollinator ratios. In addition, pollinators of sexual mimics commonly show some form of avoidance learning after attempted

copulations with sexual mimics. Such habituation can be based on the recognition of individual odor bouquets (Ayasse et al. 2000) or the avoidance of locations where sexual mimics grow (Peakall 1990; Wong and Schiestl 2002; Whitehead and Peakall 2013). Avoidance learning by pollinators probably further reduces the pollination success of sexual mimics, which may in turn fuel selection for pollinator shifts.

Pollinator shifts may sometimes also be enabled through so-called minor responders, which are often present in orchid populations (Bower 1996). Minor responders are usually much less efficient as pollinators, but may become the major pollinators when original pollinators become inefficient because of being scarcer than minor responders in particular habitats. Once a minor responder becomes more efficient that the original pollinator, there will be further selection for floral traits, typically volatile bouquets, that more efficiently attract this pollinator. Stebbins (1970) described this as passing through a stage of intermediate function. Peakall and Whitehead (2014) argued that small changes in potential pollinator attractants may be essentially neutral, and thus mutations leading to such chemical diversification may spread within populations. This scenario is probably also true for *Ophrys*, in which small amounts of compounds attractive to other pollinators are usually present (Ayasse et al. 2011). The presence of such neutral standing genetic variation in a population will speed up the potential evolutionary response to a change in pollinator availability, thus enabling quick pollinator shifts.

In this context it is interesting to ask why pollination is so specific in sexual mimics. Specificity is costly for plants, as it reduces pollination success and increases the risk of extinction were the pollinator to become rare. Would it not be preferable for plants to mimic a range of sex pheromones and thereby attract a suite of pollinators? Current evidence suggests that the mechanisms of pollinator attraction favor specificity in sexual mimics. Seemingly, odor bouquets of sexual mimics work best when they target a single pollinator species. This was shown in the thynnine pollinators of the Australian genus *Chiloglottis*, which strongly respond either to chiloglottone 1 or chiloglottone 3 as pure

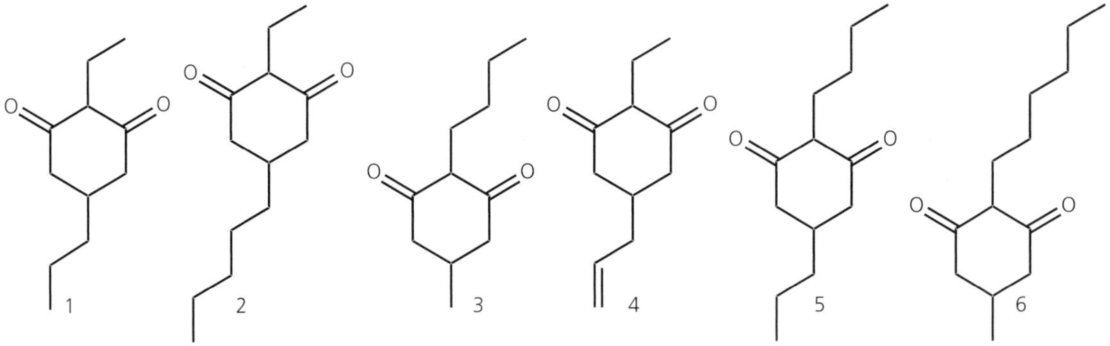

1	2	3	4	5	6
$C_{11}H_{18}O_2$	$C_{13}H_{22}O_2$	$C_{11}H_{18}O_2$	$C_{11}H_{16}O_2$	$C_{13}H_{22}O_2$	$C_{13}H_{22}O_2$
182	210	182	180	210	210

Figure 5.10 Six molecules of the "chiloglottone family" that have been identified as pollinator-attracting volatiles of Australian *Chiloglottis* orchids. Chemical formulae and molecular weight are indicated below the chemical structure. From Peakall et al. (2010).

compounds; this response, however, drops dramatically when the two chiloglottones are offered as 1:1 blends (Figs 5.10 and 5.11) (Peakall et al. 2010). A similar situation was found in two pollinator species of the European *Ophrys*: the solitary bees *Andrena nigroaenea* and *Colletes cunicularius* (Xu et al. 2012). Whereas the bees strongly responded to the odor bouquets of "their" orchids, attractiveness dropped markedly when the pollinator-attracting compounds of one orchid species were added to the odor blend of the other (Fig. 5.12). Thus both studies show that attracting two pollinators with a "two-species blend" does not attract any of the two pollinators as efficiently as would a single-species bouquet. This is likely to be a consequence of the mating behavior of the pollinators, which selects for specific responses to conspecific females in male insects, and provides a proximate explanation for pollinator specificity. In addition, though pollinator specificity tends to reduce pollination rates, it also increases pollination efficiency (the proportion of removed pollen that is deposited on a stigma) (Scopece et al. 2010). For orchid sexual mimics, a single pollinator typically forms an adaptive peak; during pollinator shifts, plants need to cross an adaptive valley with pollinator overlap (Stebbins' "stage of intermediate function"). From there, selection should favor a single pollinator species, with the consequence of

strong floral isolation in the species with the new pollinator.

An example of possible ongoing speciation by pollinator shifts is presented by Breitkopf et al. (2013). The authors of this study showed that populations of *O. sphegodes* on the west and east coasts of southern Italy are adapted to different pollinator species. Interestingly, plants in the eastern populations attract only one bee species, whereas those in the western populations attract two, with one being shared among the plant populations, suggesting that the pollinator switch is not (yet) complete. Accordingly, there is no clear neutral genetic differentiation between the populations, though the odor bouquets are different, suggesting that these populations represent different ecotypes that could be undergoing incipient speciation. Similar examples are known from the Australian hammer orchids (Menz et al. 2015).

An interesting case of a pollinator shift mediated by hybridization is presented by Vereecken et al. (2010). The authors of this study showed that F_1 hybrids of two *Ophrys* species produce a novel scent bouquet that attracts a new pollinator species. This would be a form of "saltational" pollinator shift, without a phase of pollinator overlap. However, because the two parent species have different ploidy levels—and as a consequence their hybrids are largely sterile—this pollinator shift does not lead to

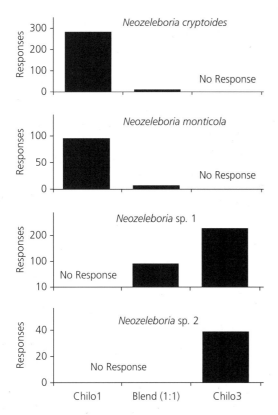

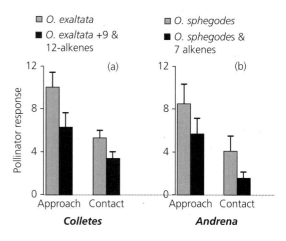

Figure 5.12 Effects of alkenes with different double-bond position on pollinator behavior, showing that additive blends of two species are less attractive than single-species blends. (a) Addition of synthetic 9- and 12-alkenes onto *Ophrys exaltata* flowers reduces their attractiveness to the pollinator, *Colletes cunicularius*. (b) Adding 7-alkenes to floral scent extracts of *Ophrys sphegodes* also reduces the attractiveness of the scent extracts to the pollinator, *Andrena nigroaenea*. Adapted from Xu et al. (2012).

Figure 5.11 Responses of four *Neozeleboria* pollinator wasp species to chiloglottones 1 and 3, and blends thereof. Chiloglottone 1 is the sex pheromone of *N. cryptoides* and the putative sex pheromone of *N. monticola*, whereas chiloglottone 3 is the putative sex pheromone of the remaining two (as yet undescribed) species. Blends of the two chiloglottones are less attractive than the pure compounds, suggesting lower pollination success of plants (hybrids) emitting both volatiles and thus selection for single-species attraction. From Peakall et al. (2010).

speciation. If the triploid hybrid population could overcome its sterility by polyploidization, however, then a new lineage could be formed. Pollinator shifts resulting from hybridization seem a plausible form of speciation, given the finding that hybridization can change odor bouquets in a non-additive way and the occasional documented hybridization within *Ophrys*.

Which traits mediate floral isolation?

To understand speciation it is useful to unravel the nature of the traits underlying reproductive isolation—assuming that isolation itself is not a trivial consequence of non-adaptive factors such as apomixis (which isolates individuals), polyploidy, and vicariance. Deciphering the genetic bases of these traits, including the number and linkage groups of genes involved, is also important for understanding the consequences of gene flow and to shed light on the developmental biology of the traits. Many studies have shown that floral isolation in sexual mimics is primarily based on differences in floral scent (Schiestl and Ayasse 2002; Peakall and Whitehead 2014; Sedeek et al. 2014). In *Ophrys* such differences lie in the relative amounts of alkenes (Xu et al. 2012), whereas in *Chiloglottis* the presence/absence of specific chiloglottones are the key parameters (Peakall et al. 2010).

By contrast, morphological traits seem to be less important for reproductive isolation. In *Ophrys*, for example, flower morphology and the direction of labellar trichomes mediate the orientation of the pollinator on the flower, and thus the positioning of pollinia on the body of the pollinators, but this barrier has been shown to be relatively "leaky" (Cortis et al. 2009). Importantly, differences in floral scent can mediate strong reproductive isolation without any changes in flower morphology. Differences in floral scent in both *Ophrys* and *Chiloglottis* are likely mediated by relatively simple mutations in regulatory genes, supporting a scenario of genic speciation (Schlüter et al. 2011; Peakall and Whitehead 2014; Sedeek et al. 2014). In genic speciation, a few genes trigger reproductive isolation and evolve to be strongly differentiated between species. The rest of the genome remains relatively little differentiated until genetic differences accumulate over time. The part of the genome not under selection may even remain porous in the face of gene flow between species and thus stay undifferentiated (Chan and Levin 2005), a situation probably found among many sexual mimics.

The possibility of speciation with gene flow

The idea that traits mediating floral isolation are based on a few simple genetic differences among species suggests that these differences can be maintained under strong selection, even in the face of gene flow. This opens the possibility of sympatric speciation without polyploidization (Dieckmann and Doebeli 1999; Schiestl and Ayasse 2002). Closely related species of sexual mimics often occur in sympatry, especially orchids of the genus *Ophrys*. Of course, sympatry may be a consequence of reproductive isolation that evolved earlier during phases of allopatry followed by range expansion. But for reasons outlined below, sympatric speciation seems possible in sexual mimics. Mathematical models suggest that speciation with gene flow is possible when (1) traits that are under divergent selection also influence reproductive isolation ("magic" traits) and (2) these traits have a relatively simple genetic basis, so recombination under gene flow does not disrupt co-adapted allele combinations. This scenario is particularly plausible in sexual mimics in which specific pollinator attraction by different scent bouquets leads to floral isolation. An obvious "magic" trait in sexual mimics is floral scent, being the major trait for pollinator attraction, but also responsible for floral isolation. Because floral scent differences probably have a simple genic rather than genomic basis (i.e., differences are based on a few genes only) (Schlüter et al 2011; Sedeeck et al 2014), they are likely to be maintained in the face of gene flow. Whereas sympatric speciation is theoretically possible in sexual mimics, this form of speciation is also notoriously

difficult to prove. First of all, orchids, the main representatives of sexual mimics, are highly unsuitable for experimental approaches and even producing flowering F_1 hybrids is difficult and time-consuming. Secondly, even though sympatric occurrences are common in many species of *Ophrys*, it is often impossible to prove that there was no allopatric phase in the past during speciation. In addition, phylogenetic reconstructions are difficult, making it challenging to identify pairs of recently diverged sister species (see Breitkopf et al. 2015 for a recent approach).

The evolution of sexual mimicry

Sexual mimicry evolved from generalized food deception, shelter pollination, or food reward systems (in *Gorteria*), as suggested by phylogenetic reconstruction of pollination systems. For example, food deception is ancestral in the lineage of *Ophrys* and allies (Inda et al. 2012), and shelter pollination is ancestral in the genus *Serapias*, in which one species has evolved sexual mimicry (Vereecken et al. 2012). Schiestl and Cozzolino (2008) have shown that pre-adaptations (exaptations) were important for the evolution of chemical mimicry in *Ophrys*. This is because the key compounds for sexual mimicry in *Ophrys*, namely alkenes, are found in flowers of several related orchid genera. Vereecken et al. (2012) provided even stronger evidence for the importance of pre-adaptations by showing that sexual mimicry has evolved from shelter pollination in two independent systems: the orchid genus *Serapias* and *Oncocyclus* irises. In shelter pollination, flowers provide chambers in which bees sleep overnight, transferring pollen through movements between shelter-providing flowers (Dafni et al. 1981). Interestingly, male insects are the primary pollinators in shelter pollination (likely because females tend to sleep in their nests), as in sexual mimicry, and similar alkenes are found in flowers of shelter pollination systems. Thus shelter pollination systems already feature several key components of sexual mimicry that can serve as pre-adaptations.

But which factors drove the evolution of sexual mimicry from other pollination systems? Vereecken et al. (2012) suggest that increased pollen transfer efficiency arising from specificity in sexual mimics was likely to be the key factor for this transition.

The evidence for this comes from another study comparing food reward, food deception, and sexual mimicry in terms of pollen transfer efficiency (Scopece et al. 2010). This study found that pollen transfer efficiency—the fraction of removed pollen that is deposited on conspecific stigmas—is much higher in sexual mimics than in the less specialized generalized food deception systems.

The mimicry system in the daisy *Gorteria diffusa* is particularly suitable for exploring the factors that favor sexual mimicry because there is a variety of forms that differ in the degree to which male bombyllid flies are induced to show copulatory behavior on them (Fig. 5.13). Ellis and Johnson (2010) also implicated male fitness as being key for the evolution of sexual mimicry because experiments showed that male flies made more interplant movements and transferred a pollen analogue more often between flowerheads than did food-seeking female flies, particularly on forms that have well-developed fly-mimicking structures.

In addition to increased pollen export, the distance of pollen flow may be a key factor driving the exploitation of male insects as pollinators in sexual mimics. Male bees fly longer distances than females even during visits to nectar-rewarding flowers (Ne'eman et al. 2006). During mating behavior, the average flight distance of male bees has been shown to be 5 m, with a maximum of 50 m (Peakall and Schiestl 2004). Importantly, the pattern of pollen flow may generally involve non-neighboring plants, in contrast to food reward systems where near-neighbor pollen transfer is the rule. In largely clonal species in particular, non-neighbor pollen flow will increase outbreeding, with likely fitness advantages (Peakall and Beattie 1996). Pollen flow among non-neighboring plants is also predicted from male mating behavior in thynnine-pollinated mimics (Peakall et al. 1990; Peakall and Beattie 1996). Whitehead and Peakall (2013) showed that flight distance after copulation attempts in a thynnine pollinator exceeded the average clone size (5 m) of the orchid *Chiloglottis trapeziformis*. Pollinators of sexual mimics rarely visit more than one flower per plant, thus a shift from generalized food deception to sexual mimicry may increase the overall efficiency of pollen export without incurring costs arising from geitonogamous self-pollination (see Chapter 2). High levels of outcrossing are therefore predicted in this pollination system, despite extensive clonality, and this has been confirmed in a recent study (Whitehead et al. 2015).

Another factor selecting for sexual mimicry may be the ability of these plants to successfully attract pollinators even at low densities (Peakall 1990). Orchids often have scattered populations with low densities of flowers that may not be able to successfully maintain visits by a reward-seeking pollinator (Nilsson 1992). However, this problem may be less severe in scent-based sexual mimicry systems. Thynnine wasps, in particular, may be able to trace and visit isolated flowers due to the long-distance pheromone communication system. Baiting experiments have largely confirmed this in several thynnine species (Menz et al. 2013).

Future directions

Despite the massive progress that has been made in understanding sexual mimicry in recent decades, many unsolved issues remain. For example, a better understanding of the biology of the models in the system has typically been difficult to achieve.

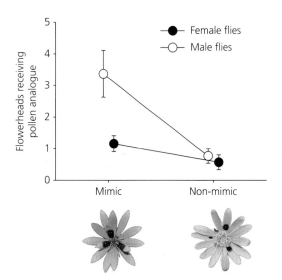

Figure 5.13 Mimetic forms of the daisy *Gorteria diffusa* outperform non-mimetic forms in terms of male pollination success (estimated from the number of flowers that receive a pollen analogue). Mimetic forms attract more male flies and these move frequently among flowerheads. Adapted from Ellis and Johnson (2010).

In Australian thynnine wasps, for example, females of the pollinator species are rarely found and very few studies have investigated their behavior, pheromone chemistry, or population dynamics. As yet it has been impossible to breed any of the pollinators of sexual mimics in the laboratory, thus preventing bioassays with naïve pollinators. As a consequence, one can never be sure about a pollinator's age or level of experience when it is used in an experiment. Naïve pollinators would, however, be necessary in order to test for innate versus acquired preferences for mating signals. It would be interesting to find out whether experienced pollinators respond less to mimics overall, suggesting some acquired avoidance of mimics during the lifetime of a pollinator. Such avoidance would suggest negative frequency-dependent dynamics between models and mimics.

It has proved difficult to estimate the local abundance, distribution, and emergence time of pollinators in their natural habitats, although these are important factors for the fecundity of sexual mimics, and may select for evolutionary responses such as pollinator shifts. Such factors are also important for assessing the potential negative fitness impacts of mimics on their pollinators. If such negative fitness impacts are widespread, then pollinators may conceivably evolve ways to avoid mimics through such means as improved learning capabilities. This could be assessed by comparing responses of local pollinators and individual insects of the same species from populations outside the range of mimics.

Sexual mimicry is an example of a very specific pollination system. Although the mechanisms of specific pollinator attraction are reasonably well understood, more insights are needed into why specificity evolves. Evidence suggests that sexual mimicry works best with highly specific mating signals, indicating that specificity is a constraint. Yet selection for specificity is also possible, for example through improved pollen export, suggesting that specificity may be adaptive and even a driver for the evolution of sexual mimicry.

A major question concerns the patterns and mechanisms of speciation in sexual mimics. In this respect, better phylogenies that resolve the relationships of closely related species, as well as between higher taxa, are badly needed. To reconcile the apparent lack of genomic differentiation with the proven strong floral isolation, we need a longer-term, spatial perspective on floral isolation, to see whether it can fluctuate with the dynamics of pollinator populations. In the "converge and diverge" scenario outlined for Darwin's finches (Grant et al. 2004), phases of strong isolation alternate with phases of gene flow. This seems possible in sexual mimics that typically lack post-pollination barriers, for example when pollinators with lower fidelity become more common and thus sharing of pollinators increases in sympatric taxa. The geographic mosaic of pollinator availability also plays a key role in pollinator shifts, but unfortunately we know very little about the local abundances of given pollinator species. Of particular interest, too, are sexual mimics with more than one pollinator, as they may offer insights into the mechanisms of multiple pollinator attraction or pollinator switching.

Sexual mimicry has evolved independently in a number of plant lineages. Therefore it would be interesting to ask which life-history and ecological conditions favor this form of pollination, and which factors constrain its evolution. An important step forward in our understanding of the evolution of sexual mimicry would be an improved grasp of the genetic background of mimicry traits in plants (see Chapter 8 for more details). For this, quantitative genetics, which requires intensive field sampling or the cultivation of plants in the greenhouse, would be necessary. For orchids in particular, improved protocols for germinating plants from seeds and growing seedlings are required in order to speed up the investigation of subsequent generations. There is no doubt that more sexual mimicry systems will be discovered in the near future; these new model systems may offer further insights and possibilities to pursue some of the outstanding questions. Of particular interest will be sexual mimics that utilize new groups of pollinators, such as the recently discovered systems involving fungus gnats, or the discovery of mimicry in new plant groups, because those have the potential to highlight general principles of this fascinating mode of pollination.

CHAPTER 6

Oviposition-site mimicry

Introduction

Of all the floral mimicry systems covered in this book, oviposition-site mimicry is probably the most widespread, in terms of geography and the number of plant species and lineages represented, and the most diverse, in terms of floral signals. It is also the most under-appreciated system of floral mimicry and is certainly less well known than floral sexual mimicry, which sometimes grabs the headlines in the popular press.

The basis of oviposition-site mimicry is that plants deploy the cues used by female insects to locate substrates for oviposition and larval development. In so doing, they deceive female insects—and sometimes also males seeking females at such sites—into serving as pollinators. The diversity of substrates used as brood sites by insects is truly staggering, providing plants with an equally wide range of ecological opportunities, or niches, for exploiting these insects.

We estimate that several thousand plant species employ some form of oviposition-site mimicry to attract pollinators. Oviposition-site mimicry has evolved in at least 23 plant families, including the Annonaceae, Apocynaceae, Araceae, Aristolochiaceae, Iridaceae, Hydnoraceae, Orchidaceae, Rafflesiaceae, Solanaceae, and Taccaceae (Jürgens et al. 2013; Jürgens and Shuttleworth 2015). While the number of insect species involved in oviposition-site mimicry systems is enormous, they appear to belong to just two orders—the Diptera (true flies) and Coleoptera (beetles).

The main evidence for oviposition-site mimicry derives from the similarity in chemical and visual signals between flowers and oviposition substrates, as well as from the behavior of insects,

including female bias among the flower visitors and egg-laying on the flowers (Urru et al. 2011). Deposition of eggs on flowers, which almost always leads to the death of the larvae, provides hard evidence that insects have been duped by the flowers. However, not all floral oviposition-site mimics induce egg-laying by duped female insects, just as not all floral sexual mimics induce ejaculation by duped male insects. The insects attracted to oviposition-site mimics are typically female, but since many insect species mate either on or in close proximity to their oviposition sites, it is not unusual for male insects to be attracted to oviposition-site mimics as well. In some cases, both sexes may simply be responding to cues indicating a food source.

Flowers that mimic oviposition sites do sometimes contain food rewards. This seeming paradox—deception combined with rewards—can be resolved by closer examination of the behavior of insects on actual oviposition sites and on the floral mimics. Flies, for example, often feed on liquid exudates of carrion or dung, and respond similarly to the nectar offered by mimic flowers. In *Stapelia* flowers the insertion of a fly's proboscis into the nectar chamber is usually necessary for the attachment of pollinaria to the mouthparts of the insect (Meve and Liede 1994). In *Rhizanthes* (Rafflesiaceae) the site of nectar production is on the tips of the perigone and thus some distance from the sexual parts of the flower (Bänziger 1996b, 2001) (Fig. 6.1). This nectar may encourage insects to further explore the flower after arriving at its edges, and it may also sustain a pollinator so that it is able to deliver pollen to distant flowers. Insects attracted to oviposition-site mimics may also feed on pollen (Gottsberger 2012). The deception in these cases relates to the

Floral Mimicry. Steven D. Johnson & Florian P. Schiestl.
© Steven D. Johnson & Florian P. Schiestl 2016. Published 2016 by Oxford University Press.
DOI 10.1093/acprof:oso/9780198732693.001.0001

Figure 6.1 A flower of the Asian parasitic plant *Rhizanthes infanticida* (Rafflesiaceae) which has attracted individuals of several different carrion fly species, including *Chrysomya* and *Lucilia* blow flies. The flies obtain nectar from pads (one is arrowed) around the margins of the flower and transfer pollen as they contact the central column of the flower. Photo © Hans Bänziger.

false promise of a brood site, not whether or not the flower offers nectar and pollen.

The term "sapromyiophily" has been used extensively in the literature on floral syndromes to denote a suite of traits in flowers that are pollinated by flies seeking rotting organic material as oviposition sites or for food (Faegri and van der Pijl 1979). In practice, this term is not particularly helpful as it has served as an umbrella for several different oviposition-site mimicry systems that are not necessarily functionally similar. It makes reference to flies, yet fails to account for the fact that many oviposition-site mimics are pollinated by both flies and beetles, and some by beetles alone, a system that has been given the rather awkward name of "saprocantharophily." Our preference has been to organize our treatment of oviposition-site mimicry according to the general properties of this form of mimicry and the different oviposition substrates (the models) that are used by insects. We identify carrion, feces, fermenting fruit, mushrooms, and aphids (and other homopterans) as the key insect oviposition sites currently known to be mimicked by flowers. Such a classification system is not perfect, however, as many flowers seem to emit signals of more than one substrate, for example both carrion and feces, and some insects use both of these as oviposition sites.

Mimicry of urine, proposed by Jürgens et al. (2006), is considered unlikely to occur in nature as

urine is not a known substrate for insect oviposition (though it is used by some adult Lepidoptera as a source of mineral nutrients, and by some other insects as a way of tracking mammal hosts). The sharing of some acids (e.g., hexanoic acid) between flowers and urine is probably a coincidence; it is more likely that the emission of these acids by certain flowers serves to attract insects that deposit eggs on decaying vertebrate tissue (Dekeirsschieter et al. 2009; Paczkowski and Schuetz 2011; Paczkowski et al. 2015).

As is the case in all mimicry systems, evolution of the traits of the mimic is advergent with respect to the model and also convergent among phylogenetically distinct mimics of the same models (Ollerton and Raguso 2006; Johnson and Jürgens 2010; Jürgens et al. 2013). However, oviposition-site mimicry differs in one important respect from other mimicry systems discussed in this book, and that is that the models are usually inanimate or dead and so cannot be negatively affected by the interaction. Therefore there is no possibility of an "arms race" between the mimics and models, as has been suggested for some other mimicry systems such as sexual deception (Lehtonen and Whitehead 2014). The benefit that mimics obtain from a higher abundance of models is not through associative conditioning of operators, as is the case in some food mimicry systems, but rather due to the positive effects of models on the reproduction of the operators. In some cases models may act as local magnets for operators, as was suggested for stinkhorn fungi that seem to benefit from growing close to badger setts with their associated badger corpses (Sleeman et al. 1997), or carrion-mimicking arums that occur on islands replete with rotting bird corpses (Stensmyr et al. 2002). On a larger scale, there are more obvious associations such as mushroom-mimicking orchids that occur in damp rainforests rich in mushrooms.

Key features of oviposition-site mimicry

Floral oviposition-site mimicry encompasses a very wide range of floral trait combinations, but there are four key associated floral features: (1) geoflory, (2) trapping devices, (3) floral gigantism, and (4) thermogenesis.

Geoflory

Flowers that mimic carrion, feces, and mushrooms are often situated at ground level, in accordance with the typical natural location of these oviposition substrates (Vogel and Martens 2000; Jürgens and Shuttleworth 2015). Insects typically walk or fly short distances onto these flowers after landing nearby (Sakai and Inoue 1999; van der Niet et al. 2011). Vogel and Martens (2000) suggested that the filiform floral appendages that characterize many oviposition-site mimics, particularly those that mimic mushrooms, may function as walkways to guide insects onto the flowers.

Geoflory is a derived condition in most plant lineages in which oviposition-site mimicry has evolved, and is therefore likely to be part of the suite of plant adaptations for this mode of deception. Examples among carrion mimics include the dead-horse arum, *Helicodiceros muscivorous*, which has an inflorescence that is angled at 90 degrees so that the spathe is often draped over the ground (see Fig. 6.4), and various stapeliads with flowers that are positioned flat on the ground (see Fig. 6.9 and Plate 7). The South African orchid *Satyrium pumilum*, which mimics carrion, is the only species in its genus that has flowers positioned at ground level (Fig. 6.9e). Once the flowers have wilted, the inflorescences elongate to disperse the seeds (van der Niet et al. 2011), suggesting that geoflory is a feature that has functional significance for pollination. In the case of the carrion mimic *Hydnora visseri*, the entrance to the trap chamber is situated at ground level and the trap chamber itself is situated below the ground (Bolin et al. 2009) (Fig. 6.2, Plate 6). Geoflory is characteristic of many parasitic plants that mimic oviposition sites, and it is thus possible that these plants with their limited above-ground structures are pre-disposed to evolve various forms of oviposition-site mimicry.

Traps and chambers

There is a strong association between oviposition-site mimicry and flowers with a chamber or trap construction (Bröderbauer et al. 2013). Chamber flowers are common in basal (early diverging) angiosperms and serve as an enclosed space in which insects brush against the reproductive parts of the flower. Trap flowers (Vogel 1965) are more sophisticated in that they detain pollinators either briefly or for extended periods of time. Flowers of this type occur in a wide range of families, including Annonaceae, Apocynaceae, Aristolochiaceae, Araceae, Hydnoraceae, Orchidaceae, and Rafflesiaceae. Traps have evolved numerous times; Bröderbauer et al. (2013) estimated that in the Araceae alone there have been at least ten independent origins of this floral mechanism.

Chambers and traps detain insects in an enclosed space for long enough for them to be coated liberally with pollen or for pollen to be transferred to the stigmas. In some cases they also force insects to exit through narrow apertures so that they can be manipulated into a precise position for pollen transfer. These functions are particularly important in oviposition-site mimicry systems because insects seeking oviposition sites tend to move around in a haphazard fashion, which does not lend itself to efficient pollen transfer. Thus chambers and traps force these insects into contact with the anthers and stigma. This is much less of an issue in food- and sex-based mimicry systems where animals assume specific positions when attempting to feed on or mate with flowers. Some oviposition-site mimics with open flowers offer nectar to entice insects into contact with the reproductive parts of the flowers. This can be seen as an alternative strategy to

Figure 6.2 *Hydnora visseri*, an African parasitic plant with trap flowers pollinated by carrion beetles. (a) The upper portion of the flower emerges above ground. (b) Cross-section of a flower (OS, osmophore; AN, anther; ST, stamen; OV, ovary). Adapted from Bolin et al. (2009). Photos © Jay Bolin. See also Plate 6.

the use of chambers and traps, although rewards are offered by some oviposition-site mimics with chamber-type flowers, possibly to keep the pollinators alive for the purposes of pollen dispersal.

The capture of insects by trap flowers is usually achieved by surfaces made slippery by downward-facing hairs and wax crystalloids. Some orchids capture pollinators by means of a hinged "see-saw" labellum (Teck 2011), or even via a self-propelled trapdoor movement of the labellum similar to that found in the trapping device of the Venus fly trap (Martos et al. 2015). Immediate escape is prevented in a number of ways. In Araceae and Aristolochiaceae there are often strategically placed trichomes that prevent exit until these trichomes lose turgidity and slough off. In *Hydnora*, entrapment of beetles is ensured by the smooth and slippery vertical surface of the wall chamber (Marloth 1907; Bolin et al. 2009). Beetles can only escape once this surface becomes textured on the third day after anthesis. In some trap flowers insects leave via a new exit formed by a secondary opening (Vogel 1978). In orchids, insects are not detained in the labellum chamber for any length of time and are free to escape immediately, although it may take them up to 30 minutes to extricate themselves via an exit tunnel that brings them into contact with the reproductive parts of the flower (Plate 8).

Bröderbauer et al. (2012) found a significant association between fly pollination and traps in the Araceae. This association is likely because oviposition-site mimicry systems are dominated by fly pollination. The hairs and sterile flowers that block the exit from traps in some Araceae genera may also be ineffectual against sturdier insects such as beetles.

Trap flowers, especially those that detain insects for extended periods, are usually protogynous. This means that insects arrive with pollen that is immediately deposited on receptive stigmas in the female stage, and then leave with a fresh load of pollen once the flower has reached a male stage. In the dioecious genus *Arisaema* (Araceae) male flowers provide an escape route for insects but female flowers have no exit route and insects therefore perish inside (Vogel and Martens 2000). Trap flowers that detain insects only momentarily—so-called imperfect traps or semi-traps—often accomplish pollination and pollen deposition together as the insect exits the flower. These types of traps are found in slipper orchids that have a smear of pollen that has to be forcibly applied to or removed from the bodies of insects (Bänziger et al. 2012). In other orchids, a hinged labellum catapults insects against the rostellum, resulting in the attachment or removal of pollinaria (Teck 2011).

The potential for pollen transfer among trap flowers can be investigated by labeling insects in trap flowers with a suitable marker, such as a dab of paint, so that their movements between plants can be tracked. Using this approach, Beath (1996) found that individuals of the carrion beetle *Phaeochrous amplus* (Hybosoridae) moved up to 37 m between the inflorescences of *Amorphophallus johnsonii* (Araceae) (Fig. 6.3). He described this arum as having a "powerful aroma of rotting fish and faeces." Anthesis lasts for 24 hours and commences in the evening when scent production is at a maximum and beetles covered in pollen fall into the lower spathe, which has a slippery texture that prevents escape. The beetles depart by crawling up the spadix the following evening when pollen is produced and the odor has decreased markedly.

Bolin et al. (2009) labeled hide beetles (*Dermestes maculata*) entering the trap flowers of *Hydnora africana* to establish the timing of their escape. They found that beetles are imprisoned during the pistillate (female) phase and that at least 55% of the marked individuals had managed to escape by the third day after pollen is shed.

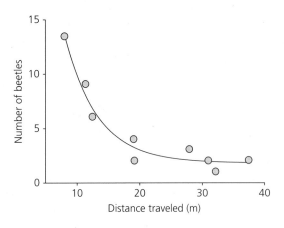

Figure 6.3 The dispersal of marked carrion beetles among inflorescences of the arum lily *Amorphophallus johnsonii* in Ghana. Adapted from Beath (1996).

Thermogenesis

There is an association between thermogenesis (the production of sufficient heat to raise tissue temperature above that of ambient temperature) and oviposition-site mimicry in plants, albeit a loose one. Thermogenesis is also known from conventionally rewarding beetle-pollinated angiosperms and even from cycads that have brood-site mutualisms with beetles (Seymour and Schultze-Motel 1997). There are at least 12 angiosperm families in which thermogenesis has been recorded, including the early diverging basal woody (Annonaceae and Magnoliaceae) and herbaceous angiosperms (Aristolochiaceae and Nymphaeaceae) and primitive monocots (Araceae, Arecaceae, and Cyclanthaceae), as well as the eudicot lotus family (Nelumbonaceae) and some parasitic lineages (Hydnoraceae, Rafflesiaceae). Thermogenic species tend to have large chamber flowers and are usually strongly scented. Pollination of thermogenic species is mostly by beetles or, to a lesser extent, flies.

The temperature of thermogenic flowers can be raised by as much as 35°C above ambient temperature (Seymour and Schultze-Motel 1997). Heat production in both plants and animals is due to the mitochondrial process of aerobic respiration. The mechanism in plants was thought to involve mainly the dissipation of redox potential by mitochondrial alternative oxidase (AOX), but there is now some evidence that uncoupling proteins (UCP) that serve to dissipate mitochondrial electrochemical potential may also be involved (Onda et al. 2008). Both AOX and UCP genes are expressed during thermogenesis in some arums (Onda et al. 2008), but a recent analysis of the transcriptome of the non-thermoregulatory species *Arum concinnatum* showed expression of AOX-encoding genes, and not UCP-encoding genes, during the heating phase (Onda et al. 2015). Some plants exhibit increasing respiration rates (and therefore heat production) when there is a drop in ambient and tissue temperatures (Seymour and Schultze-Motel 1997). This represents a form of thermoregulation that is not yet fully understood at the molecular level.

Mass-specific rates of respiration for the thermogenic male florets of arums are similar to the highest rates of respiration recorded for flying insects and birds (Seymour 2010). An important limitation of respiration in flowers is the supply of oxygen, as flowers lack a convective system for gas transport. Experimental data suggest that heat production by some flowers is indeed limited by the supply of oxygen through the intercellular diffusion pathway to the mitochondria (Seymour et al. 2015). This constraint would be reduced in less dense tissue, and this may explain the evolution of large spongy flowers such as those of *Rafflesia* and the hollow pith of the appendix in arum lilies (Seymour 2010).

Floral thermogenesis may have more than one function, even in the same plant. For example, in aroids there are often several episodes of temperature increase including heating of the appendix, followed by heating of the male florets (Seymour et al. 2009). There are a number of hypotheses that have been put forward to explain the evolution of floral thermogenesis (Seymour and Schultze-Motel 1997), and only some of these relate to mimicry. These hypotheses include: (1) increased volatilization of scent (as production of scent often coincides with production of heat), (2) a heat reward for insects, (3) promotion of insect activity inside the chamber, (4) a temperature cue that insects associate with rotting plant or animal material, and (5) promotion of pollen and pollen tube development (Seymour et al. 2009).

Experimental investigations of the function of additional heat in flowers are scarce; however, some insights have been gained from studies carried out on the thermogenic dead-horse arum, *Helicodiceros muscivorous* (Stensmyr et al. 2002; Seymour et al. 2003; Angioy et al. 2004) (Plate 6). This species produces heat and scent from the appendix on the first day of anthesis, when it attracts flies searching for carrion oviposition sites, but on the second day heat and scent production by the appendix cease and flies depart from the inflorescences (Stensmyr et al. 2002; Seymour et al. 2003; Angioy et al. 2004). In addition, the male florets are highly thermogenic during the night that flies are trapped within the floral chamber (Seymour et al. 2003) (Plate 6). Attractiveness of the inflorescences to flies could be restored on the second day by application of a mixture of artificial heating of the appendix by means of a coiled resistance wire and scent (applied in a cotton wick). Scent alone was not as attractive as the combination of heat and scent. In this study attractiveness was measured as the percentage of flies that land on the appendix, because flies landing on the appendix are much more likely to be trapped than those landing on the spathe. The temperatures of gull carcasses in the vicinity of the arums were found to be elevated about 12°C above ambient, which is similar to that recorded for the arum inflorescences on the first day of anthesis and highlights the possibility that heat production could be a "mimicry" trait.

Although there is some evidence that heating of the appendix may encourage flies to move into the chamber and/or the volatilization of scent compounds, the heating phase of the male florets in *H. muscivorus* actually extends throughout the night (Fig. 6.4). This heating phase cannot serve for the attraction of flies and is more likely to be related to the maturation and release of pollen (Seymour et al. 2003, 2009). The idea that heating itself stimulates fly activity has been rejected on the grounds that by early morning flies in the traps are sluggish on account of their circadian rhythms and perhaps even anesthetized by high levels of CO_2 in the chamber (Seymour et al. 2003). Studies of other *Arum* species have failed to show that heat itself is attractive to insects (Kite et al. 1998).

Floral gigantism

There is an intriguing association between floral gigantism and oviposition-site mimicry, particularly carrion mimicry. The carrion mimics *Rafflesia arnoldii*

(Rafflesiaceae) and *Amorphophallus titanum* (Araceae) have the distinction of producing the world's largest flower and blossom (a flower-like inflorescence), respectively. Flowers of *R. arnoldii* can measure 1 m in diameter, while the blossom of *A. titanum* can be nearly 3 m tall. Other carrion mimics that have exceptionally large flowers include *Aristolochia grandiflora* (Aristolochiaceae), *Stapelia gigantea* (Apocynaceae), and *Hydnora africana* (Hydnoraceae).

The evolution of giant flowers has occurred at an accelerating rate in the Rafflesiaceae (Davis et al. 2007; Barkman et al. 2008). This family represents a lineage that evolved within the broader Euphorbiaceae clade; floral size evolution along the stem of this lineage has been 91 times faster than in lineages of Euphorbiaceae (a family known for conservative evolution of small flowers) or even other lineages within the Rafflesiaceae (Fig. 6.5). Flower diameter increased from 24 to 189 mm over a period of 46 million years, representing an 8% increase in flower size per million years (Davis et al. 2007). More detailed phylogenetic analyses have shown that the genus *Rafflesia* is only 12 million years old, with much of the diversification in this group occurring in the past 1–2 million years (Barkman et al. 2008). The rate of flower size evolution in *Rafflesia* is 20 times higher than in the rest of the family, and there are examples of recently diverged sister species (e.g., *Rafflesia pricei* and *Rafflesia keithii*) that

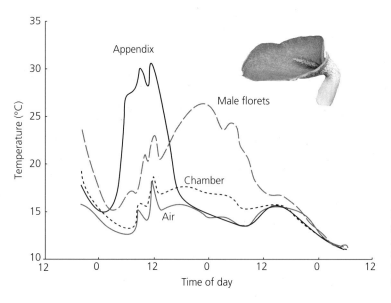

Figure 6.4 The time course of temperature changes in parts of an inflorescence of *Helicodiceros muscivorous* under field conditions. Photo © Roger Seymour and Marc Gibernau. Adapted from Seymour et al. (2003).

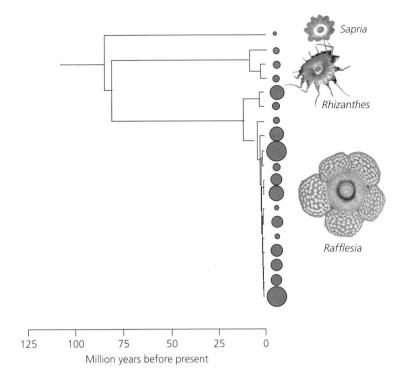

Figure 6.5 The evolution of giant flowers occurred particularly rapidly within the Rafflesiaceae. The highest rates have occurred within the genus *Rafflesia* itself; some sister species in *Rafflesia* have evolved a three-fold difference in flower size in less than 1.5 million years. Circles represent the relative flower sizes of different species, up to a maximum diameter of 1 m. Images: *Sapria poilanei* © Hans Bänziger, *Rhizanthes infanticida* © Hans Bänziger, *Rafflesia cantleyi* © Suk-Ling Wee. Adapted from Barkman et al. (2008).

have more than three-fold differences in flower size (Fig. 6.5). Given these rates of evolution it seems likely that flower size is under particularly strong selection in this group. Although the overall trend is for the evolution of larger flowers, there may also have been reversals from larger to smaller flowers.

It is not yet completely clear why floral gigantism is more pronounced among carrion mimics than among flowers that mimic other oviposition substrates. Explanations that have been put forward include visual mimicry of large animal carcasses, facilitation of heat production, and emission of large amounts of scent consistent with that produced by large dead animals. The importance of emission of large amounts of scent for attraction of blow flies such as *Lucilia sericata* has been shown by experiments in which fly attraction increased strongly with increasing emission of the volatile compound dimethyl trisulfide (DMTS) (Fig. 6.8) (Brodie et al. 2016). This compound is a major component of the floral scent of carrion mimics with giant flowers.

Davis et al. (2008) suggested that floral gigantism may arise through a coevolutionary arms race between flowers and carrion flies in which flowers are selected to exceed a certain threshold in size and flies are selected to avoid flowers below a certain threshold because of the negative effects of being deceived. We think that this scenario is unlikely; firstly because diversion by carrion mimics is unlikely to have much general impact on the fitness of flies in natural ecosystems, and secondly because animal carrion is variable in size and it would not benefit flies to ignore signals associated with small carrion, particularly if large carrion were not available. Flies and beetles are likely to respond in greater numbers (and from a greater distance) to flowers with a particularly strong emission of the olfactory signals of carrion, and this would select for large flowers that can emit larger amounts of scent (and are also able to produce and maintain heat). The largest carrion mimics (*Rafflesia* and *Amorphophallus* species) occur at very low densities in tropical forests, which is consistent with the distance of detection hypothesis. Yet there are some carrion mimics that have small flowers (van der Niet et al. 2011), and this suggests that the evolution of floral gigantism is dependent on the environmental context or is evolutionarily constrained in some lineages.

Carrion mimicry

Carrion flowers mimic the odor and appearance of animal carcasses for the purpose of attracting necrophagous (carrion-feeding) insects that use animal corpses as brood sites. There is, however, a continuum between flowers that mimic carrion and those that mimic feces. Some plant species (and some stinkhorn fungi) are not easily characterized as being either carrion or feces mimics as they emit signals of both these substrates (Fig. 6.6). Indeed, some insects have larvae that feed on both carrion and feces, and there are some key volatile compounds that are emitted by both these substrates. Further

complicating the assessment of carrion and fecal oviposition-site mimicry is that many adult flies use both these substrates as a source of protein-rich food. Some individuals of a species may visit either carrion or feces for food, while others, particularly gravid females, will be seeking oviposition sites (Bänziger and Pape 2004; Brodie et al. 2016). It should therefore be acknowledged that the evolution of both carrion and fecal mimicry may be influenced by additional fitness benefits obtained from attracting insects that seek food.

Broad-spectrum floral mimicry of both carrion and feces may be ecologically viable because insects attracted to carrion are often similar

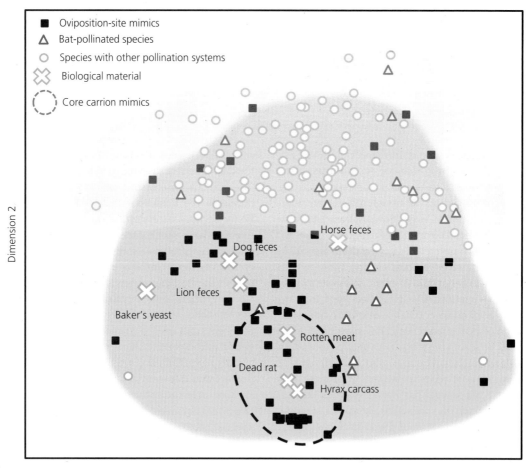

Figure 6.6 Representation of the degree of similarity in volatile emissions between flowers of 172 plant species and various oviposition sites used by insects. The distance between symbols represents the degree of dissimilarity in the overall chemical composition of scent. Plant species circled with a broken line emit scent which is very similar overall to that emitted by decaying animal flesh; these plants are either known or predicted to be carrion mimics. Adapted from Jürgens et al. (2013).

morphologically to those that feed on feces, and thus are functionally similar in terms of the mechanics of pollen dispersal. If there is no trade-off in attractiveness for a plant that emits both carrion and feces signals then the consequence will be that it will gain the advantage of attracting a greater number of (functionally equivalent) pollinators than would be the case if it emitted the signals of only one type of oviposition substrate. However, at least one study has shown that a combination of carrion and fecal odors may actually repel gravid female blow flies (Brodie et al. 2016), suggesting that trade-offs leading to specialization may occur. There is also evidence that a great many plant species are specialized for pollination by necrophagous insects, and these can be considered to be true carrion mimics (Jürgens et al. 2013). Some are even specialized for pollination by a particular necrophagous insect group. For this reason we have opted to separate our discussions of carrion and fecal mimicry, while acknowledging that many plants produce flowers that combine the signals of both carrion and feces.

There is a wide range of fly and beetle taxa that use rotting animal flesh as a substrate for larval development. These include blow flies (Calliphoridae), flesh flies (Sarcophagidae), house flies (Muscidae), and a large number of beetle groups, including hide beetles (Dermestidae) and carrion beetles (Silphidae). Female carrion insects seek out animal corpses on which to lay their eggs. Male carrion insects are often also attracted to corpses, as they mate with females either on the corpse itself or close to it. Animal corpses go through different stages of decay, and there is niche partitioning of insects according to both the size of carrion and the stage of decay. Blow flies, for example, are often attracted to fresh animal corpses, while carrion beetles tend to be attracted to corpses in a later stage of decay. There is now good evidence that insects use volatile signals to assess the stage of degradation of animal carcasses (Brodie et al. 2016). Niche partitioning among necrophagous insects, therefore, has important implications for the evolution of signals in floral carrion mimics.

The importance of oligosulfides

Flowers that specifically mimic rotting animal flesh have an unmistakable "cadaverous" odor.

Sprengel (1793) correctly identified the function of this odor in the South African succulent *Stapelia hirsuta* as being to "lure bottle and carrion flies to which the stench is highly agreeable, and seduce them to fertilize the flower" (translation by Vogel 1996). These early remarks by Sprengel represent, to the best of our knowledge, the first mention of floral mimicry in any plant. The chemical similarity between the floral odor of carrion mimics and actual animal carrion was only established some 200 years after the publication of Sprengel's book (Kite and Hetterscheid 1997; Stensmyr et al. 2002). These flowers invariably emit various oligosulfides (dimethyl mono-, di-, and trisulfides) identical to those emitted by carrion (Stensmyr et al. 2002; Johnson and Jürgens 2010; Urru et al. 2011; Jürgens et al. 2013). These oligosulfides are derived from bacterial degradation of protein, specifically the sulfur-containing amino acids cysteine and methionine (Paczkowski and Schuetz 2011). Many studies have shown that flies and beetles use these oligosulfides as cues to locate carrion (Urru et al. 2011). The most important compound appears to be DMTS which elicits physiological antennal responses and positive behavioral responses in a very wide range of carrion insects (Brodie et al. 2014; Zito et al. 2014). However, dimethyl disulfide (DMDS), which is often co-present with DMTS, also functions as an attractant in some systems (Kalinova et al. 2009).

The role of DMDS and DMTS in floral carrion mimicry was demonstrated in an elegant study of the Mediterranean arum *Helicodiceros muscivorus* by Stensmyr et al. (2002). Blow flies (Calliphoridae) are attracted to the inflorescences on the first day of anthesis which coincides with the female floral phase and maximum emission of scent dominated by oligosulfides. Flies are trapped within a chamber, but are released with a fresh load of pollen on the second day of anthesis when the scent is barely noticeable and the inflorescence is in its male phase. However, experimental addition of cotton wicks soaked in DMDS and DMTS to inflorescences on the second day of anthesis restores the attractiveness of the inflorescences to flies. Antennae of the blow flies respond almost identically to the scent of the arum and the scent of a dead animal, suggesting close chemical mimicry by the arum. Furthermore, the behavior of the flies in this system strongly suggests that flies are unable to discriminate between the floral mimic and actual carrion.

The central importance of oligosulfides for the evolution of carrion mimicry systems has been further highlighted in studies of the South Africa hyacinth genus *Eucomis*. Some *Eucomis* species are pollinated by carrion flies and emit oligosulfides, while others are pollinated by spider-hunting wasps and do not emit these compounds. The dimensions and visual signals of flowers of wasp- and fly-pollinated *Eucomis* species are scarcely different. If open vials containing a mixture of the oligosulfides DMDS and DMTS are placed at the base of the flowering stem of one of the wasp-pollinated species, carrion flies are attracted and effectively pollinate the flowers (Fig. 6.7) (Shuttleworth and Johnson 2010). This demonstrates that a change in biochemical pathways resulting in emission of oligosulfides would be sufficient to induce a shift from pollination by wasps to pollination primarily by carrion flies.

In general, emission of oligosulfides is closely associated with pollination by carrion flies and beetles (Urru et al. 2011). Biochemical convergence in oligosulfide emission is evident in a wide range of unrelated plants that are pollinated by carrion insects (Fig. 6.6) (Jürgens et al. 2013). Interestingly, oligosulfides are also known as a component of the floral scent of New World plants pollinated by bats (Knudsen and Tollsten 1995). These bat-pollinated flowers are usually situated high in the canopy, which is a context that flies do not appear to associate with carrion.

Visual and tactile imitation of carrion

Many carrion flowers bear an uncanny resemblance to animal carcasses, in being brown or pink with purple blotches and covered in hairs much like the skin of a mammal (Fig. 6.9, Plates 6 and 7). Darker central parts of the flowers often resemble open wounds or other potential entry points into the carcass. Some biologists have even likened the appearance of the dead-horse arum (Fig. 6.4, Plate 6) to the rump of a dead mammal, complete with an "anus" and a "tail" (Seymour et al. 2003).

Studies on the responses of necrophagous insects to various cues are motivated mainly by the importance of these insects for agriculture, human health, and forensic science. Here we extract some

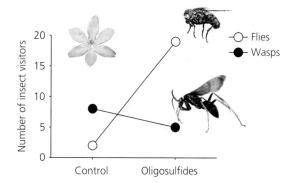

Figure 6.7 Addition of oligosulfides to inflorescences of the wasp-pollinated species *Eucomis comosa* results in attraction of carrion flies which then pollinate the flowers. Adapted from Shuttleworth and Johnson (2010). Photos © Adam Shuttleworth and Steven Johnson.

findings from these studies that shed light on the possible functional significance of the traits of carrion flowers.

In laboratory experiments using inverted cups laced with DMTS as the primary attractant, gravid females of the common green blow fly *Lucilia sericata* preferred to settle on cups covered with reddish-brown cheesecloth than on cups covered with white cheesecloth (Brodie et al. 2014). This preference became greater as the concentration of DMTS was increased (Fig. 6.8). The preference for darker cues (red or black) over lighter cues (white or yellow) was also demonstrated in a companion set of field experiments. In a similar set of field trials conducted in Norway, female *Lucilia* and *Calliphora* blow flies were strongly attracted to black lures in the context of a carrion odor, but this effect was not evident for males (Aak and Knudsen 2011). Wall and Fisher (2001) found that 6-day-old female *Lucilia sericata* blow flies preferentially landed on a 15 cm × 15 cm black square (superimposed on a white background) when a carrion odor source was in close proximity (<30 cm), but that when the odor source was placed 60 cm from the black square, flies preferentially landed on the white background near the odor source. This attraction to dark objects may allow female blow flies to locate small carrion on a lighter background and darker wounds and orifices on paler carrion. Blow flies also have a penchant for entering dark cavities in animal carcasses, though there is evidence that some blow fly species actually avoid wounded areas or natural cavities such as the eyes or anus when laying eggs (Charabidze

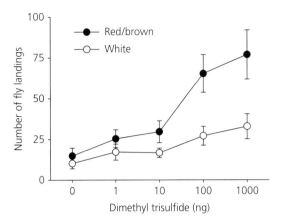

Figure 6.8 Landing responses of the blow fly *Lucilia sericata* are influenced by oligosulfides and the spectral properties of the target. Landings on a dark-colored target increase disproportionately as the concentration of dimethyl trisulfide is increased. Adapted from Brodie et al. (2014).

et al. 2015). In general, there is now good evidence that the combination of DMTS and a dark color represents a bimodal cue that signals a suitable oviposition site for gravid female blow flies. This has important implications for the evolution of the color and patterning of carrion flowers.

More et al. (2013) tested the responses of flies to white versus black model flowers in the presence and absence of a mixture of DMDS and DMTS. As expected, oligosulfides greatly increased the attractiveness of the model flowers to flies. Flies tended to land on the models even when the odor source was placed several centimeters away. Contrary to the studies described above, more flies landed on white than on black models, but the sex of the flies was not recorded in this experiment and not all the flies attracted were necessarily searching for oviposition sites. Adult blow flies of both sexes, for example, feed on flowers and can be attracted to lighter-colored traps.

A clear-cut case for the functional role of darker colors in a carrion mimicry system was presented by Chen et al. (2015). These authors extracted the dull red pigments from the inflorescence of the Asian carrion mimic *Amorphophallus konjac* (Araceae) and then added these pigments to artificial inflorescences made of filter paper. Plain white filter paper served as a control. Carrion insects (Calliphoridae, Sarcophagidae, and Muscidae) were attracted more to the dull red artificial inflorescences than to the white ones, and this effect was much stronger

in terms of the absolute number of flies attracted when the artificial inflorescences emitted a mixture of oligosulfides. The attraction of a few flies to the unscented artificial inflorescences in this study may have been due to the influence of nearby (2 m away) inflorescences that emitted oligosulfides. These experiments underline the importance of multimodal cues for the attraction of carrion insects.

The ultimate deception of carrion insects is evident when insects lay their eggs, or in the case of some flesh flies (Sarcophagidae) their live larvae, on the flowers of deceptive species. Though this behavior has been reported in insects visiting a wide range of carrion flowers, it is by no means certain that they will lay their eggs on the flowers, even when it is obvious that they have mistaken them for oviposition sites. Gravid female blow flies attracted to the combination of DMTS and a dark-colored object do not usually begin laying eggs immediately. The question that arises, then, is what are the additional cues that induce egg-laying in insects, and how does the possession or existence of these cues enhance the fitness of the plant?

Blow flies have been shown to use a range of sensilla on their antennae, labellum, and ovipositor prior to oviposition, suggesting the importance of tactile cues for these insects. The use of tactile sensilla is evident from observations that blow flies probe with their ovipositor before laying their eggs (Wallis 1962; Bänziger 1996b) and in experiments in which the flies tended to lay their eggs in cavities, despite being deprived of localized odor cues (Wallis 1962). In turn, the experimental application of wax to the flies' anal leaflets or ovipositor drastically altered their egg-laying behavior and rendered them unable to locate an egg-laying medium through holes drilled in a Petri dish (Wallis 1962).

Bänziger (2001) argued that the dense tangle of soft hairs on the perigone of *Rhizanthes infanticida* acts as a tactile cue stimulating oviposition behavior in blow flies. Up to a thousand fly eggs have been observed on a single flower of this species, while no eggs have been observed on flowers of related species such as *Rhizanthes deceptor* that lack densely set soft hairs. In a personal communication, Hans Bänziger described his experiments in Thailand with sylvatic blow flies (those that occur in or affect wild animals) such as *Chrysomya villeneuvi* and *Lucilia porphyrina*. He found that the flies prefer to oviposit

Figure 6.9 Convergent evolution of mottled patterning among species with flowers that are pollinated by carrion flies: (a) *Aristolochia cymbifera*, © Rogan Roth; (b) *Rafflesia cantleyi*, © Suk-Ling Wee; (c) *Orbea variegata*, © Adam Shuttleworth; (d) *Huernia hystrix*, © Steven Johnson; (e) *Satyrium pumilum*, © Steven Johnson; (f) *Ferraria crispa*, © Steven Johnson. Images are not to scale. See also Plate 7.

in the deep fur of animal carcasses, even when offered an alternative option, in this case rotting liver. His explanation is that this protects the flies and their larvae from wasp and tiger beetle predators and from ants that prey on the eggs. However, some temperate blow flies are known to lay eggs on substrates that are not hairy, such as rotting liver. Sheep blow flies (*Lucilia sericata*), for example, prefer to lay eggs on those parts of rat cadavers on which the hair is short, rather than on parts where the hair is long (Charabidze et al. 2015).

Are there specialized niches for carrion mimicry?

We began our discussion of carrion mimicry with the caveat that some flowers appear to mimic the signals of both carrion and feces and thus represent a more generalized form of oviposition-site mimicry. Is there any evidence for an evolutionary trend in the opposite direction, i.e., that flowers can mimic a particular type of carrion and thus represent a highly specialized form of oviposition-site mimicry? Necrophagous insects are known to partition the general carrion niche into various sub-niches according to carrion size, type, and degree of decomposition, thus making it plausible that flowers could exploit one of these sub-niches by emitting signals that insects use to identify their favored carrion oviposition sites.

Blow flies are usually the first insects attracted to a newly dead animal but they avoid corpses in later stages of degradation, probably because the larvae of flies that arrive later face intense competition (and even cannibalism) from the already-hatched larvae of flies that arrived earlier. It is well known that the oligosulfide DMTS is a key attractant to female blow flies and that DMTS is emitted by corpses even in later stages of degradation, so how do flies avoid mistakenly laying eggs on older carcasses? A recent study suggests that gravid female blow flies may use the compound indole, which is emitted mainly during later stages of degradation, as a cue to avoid older carrion (Brodie et al. 2016). This idea is based on an experiment in which gravid female flies were attracted less to a mixture of carrion and indole than to carrion alone (Fig. 6.10). Indole is also a major component of the odor of feces (Johnson and Jürgens 2010; Brodie et al. 2016). These findings suggest that specialized carrion mimicry may evolve because there is a trade-off between the emission of carrion odors and fecal odors. Such trade-offs are one of the important drivers of the evolution of floral specialization (Aigner 2001).

Some studies of floral carrion mimics have shown that they attract an assemblage of flies that is similar to that found on nearby animal carcasses. However, there is also evidence that carrion mimics can specialize on a particular component of the local

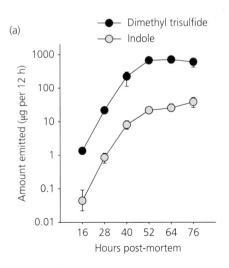
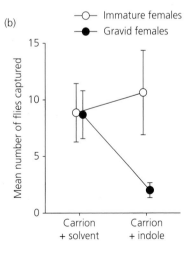

Figure 6.10 The influence of indole on attraction of female *Lucilia sericata* blow flies. (a) The amounts of dimethyl trisulfide and indole emitted increase with the age of a carcass. (b) Addition of indole to carrion has no influence on attraction of immature females that feed on animal feces, but deters gravid females that lay eggs on fresh animal carrion. Adapted from Brodie et al. (2016).

carrion fly community (Bänziger and Pape 2004). A particularly clear case involves the South African orchid *Satyrium pumilum*, which is pollinated exclusively by flesh flies (Sarcophagidae) (van der Niet et al. 2011). These flies appear to be unable to discriminate between the orchid flowers and actual carrion, and sometimes deposit live larvae on the orchid. Blow flies, on the other hand, were common on nearby animal carcasses yet strikingly absent from the pollinator assemblage recorded from the orchid flowers. How does the orchid seemingly specialize for pollination by a particular component of the local carrion fly assemblage? Its scent is composed of a mixture of oligosulfides, 2-heptanone, *p*-cresol, and indole, and is at least 50-fold weaker than that emitted by small items of carrion. One possibility is that blow flies perceive the orchid flowers as representing carrion which is too small for oviposition. It has been suggested that flesh flies are not deterred by small carrion because they deposit live larvae which are at a more advanced stage of development (Denno and Cothran 1975). However, the idea that carrion size is an important niche axis for flies remains controversial (Kuusela and Hanski 1982). One alternative possibility for the exclusive attraction of flesh flies to *S. pumilum* is that the chemical composition of its scent is repellent to blow flies, particularly given that it has been shown that gravid females of some species are deterred by the combination of DMTS and indole (Fig. 6.10). Yet another possibility is that sarcophagids are attracted by DMDS, which dominates the floral scent of this orchid, while the amounts of DMTS are too small to warrant investigation by calliphorids.

Animal carcasses in late stages of degradation are attractive to a range of specialized carrion beetles that feed on dried connective tissue and dry flesh and skin (von Hoermann et al. 2013). This particular carrion niche is exploited by several plant species including members of *Dracunculus* and *Amorphophallus* (Araceae) and *Hydnora* (Hydnoraceae) (Beath 1996; Bolin et al. 2009). The trap blossoms of *Hydnora africana* and the closely related congener *H. visseri*, which appear just above ground level, are pollinated mainly by carrion beetles belonging to the genus *Dermestes* (Dermestidae) (Marloth 1907; Bolin et al. 2009). The scent of *H. africana* consists of a rich mixture of oligosulfides and various acids

(Burger et al. 1988). The latter are typical of animal carcasses in later stages of degradation (Dekeirsschieter et al. 2009).

Fecal mimicry

Animal feces serve as a protein-rich substrate for the larvae of many fly and beetle taxa. There are usually differences in the chemical composition of herbivore feces, which is often dominated by *p*-cresol and various terpenoids, and carnivore and omnivore feces, which tends to contain higher amounts of phenol, oligosulfides, and indole (Johnson and Jürgens 2010; Midgley et al. 2015). Flies show specialization for particular types of feces. For example, the Asian fly species *Sarcophaga albiceps* will naturally oviposit on carnivore and omnivore feces but not on herbivore dung; if larvae are experimentally transferred to the latter they do not survive (Bänziger and Pape 2004). Carrion and feces have many volatile compounds in common (Jürgens et al. 2013) and some "coprophagous" species can also complete their larval development on a carrion substrate (Bänziger and Pape 2004). However, carrion and feces usually differ strongly in quantitative aspects of their chemistry—the scent of carrion tends to be dominated by oligosulfides whereas that of dung is usually dominated by aromatics (phenol and *p*-cresol) and nitrogen-containing compounds (skatole and indole).

Plant species with a floral fragrance characterized by dung volatiles occur in many plant families, including the Apocynaceae, Araceae, Orchidaceae, Lowiaceae, Sterculiaceae, and Aristolochiaceae (Young 1984; Kite et al. 1998; Sakai and Inoue 1999; Jürgens et al. 2006; Ollerton and Raguso 2006; Johnson and Jürgens 2010; Urru et al. 2011). Unfortunately, pollinators have been recorded for relatively few of the species that fit the chemical profile of dung mimicry.

One of the best-studied systems of dung mimicry occurs in the European species *Arum maculatum* (Kite 1995; Kite et al. 1998; Chartier et al. 2013). This species is pollinated primarily by individual females of the drain fly *Psychoda phalaenoides* (Psychodidae), which breeds in dung. The inflorescence scent contains compounds such as indole and *p*-cresol that are well-known components of the scent of herbivore dung. Field bioassays have shown that

Psychoda flies are particularly strongly attracted to *p*-cresol; however, a mixture of *p*-cresol, indole, and 2-heptanone is more attractive to these flies than any of these compounds individually. The inflorescence is thermogenic, but experiments have failed to establish that the flies are attracted to heat (Kite et al. 1998). The related species *Arum italicum* is also pollinated mainly by *Psychoda* flies but has a more generalist pollination system and also attracts female flies belonging to other families (Albre and Gibernau 2008; Chartier et al. 2013).

Pollination by coprophagous (dung-feeding) scarab beetles has evolved in several lineages of the Araceae, including *Amorphophallus*, *Arum*, and *Sauromatum* (Kite et al. 1998; Punekar and Kumaran 2010). There is a strong evolutionary association between emission of indole and skatole (both emitted by animal feces) and pollination by dung beetles in the Araceae (Schiestl and Dötterl 2012).

A remarkable system of pollination by dung beetles has been recorded in the Asian genus *Orchidantha* (Zingiberales: Lowiaceae). Detailed investigations of *Orchidantha inouei* in Borneo revealed that at least four species of *Onthophagus* dung beetles (Scarabaeidae) pollinate this species (Sakai and Inoue 1999). *Orchidantha* produces bizarre flowers which are dark purple with a fetid odor and situated at ground level. Pollen is transferred to the beetles when they crawl onto the labellum and beneath the lateral petals where the stigmas and anthers are hidden (Fig. 6.11). Unlike other members of the Zingiberales, *Orchidantha* flowers lack nectaries. *Onthophagus* beetles are known to dig a tunnel beneath animal dung (and occasionally also carrion) and to transport dung to the base of the tunnel. The four species that pollinate *O. inouei* have also been captured on both dung and carrion bait. Although the scent of *O. inouei* has not, to our knowledge, been analyzed, DMDS, *p*-cresol, and indole, components of the scent of dung and carrion, have been found to be emitted by flowers of a related species, *Orchidantha fimbriata* (Feulner et al. 2015).

Fermenting fruit mimicry

A large number of beetle and fly taxa use fruit and other sugar-rich organic matter as larval brood sites. They locate these brood sites primarily using

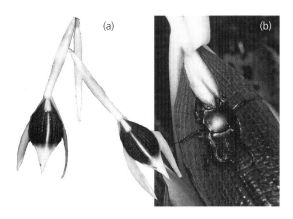

Figure 6.11 The Asian plant *Orchidantha inouei* deploys mimicry of dung to attract beetle pollinators. (a) Flowers of *O. inouei* are displayed just above the ground. Photo © Hidetoshi Nagamasu. (b) A *Paragymnopleurus pauliana* dung beetle inserting its head under the lateral petals. Photo © Anne Sakai.

chemical cues. These are typically volatiles emitted by yeasts and bacteria that inhabit fruits, and possibly also volatiles produced by fruits to attract vertebrate dispersal agents. Flowers that emit similar blends of volatiles can attract these beetles and flies as pollinators.

Fruit mimicry is found in some basal angiosperms. A large number of small-flowered tropical Annonaceae are pollinated by sap beetles (Nitidulidae), particularly of the subfamily Carpophilinae ("fruit-lovers") which are closely associated with rotting fruit. Examples of this form of oviposition-site mimicry are found in species-rich genera such as *Guatteria* and small-flowered members of *Annona* and *Duguetia* (Fig. 6.12). These tend to emit a fruity, often banana-like, odor composed of fruit esters and alcohols (Goodrich et al. 2006; Goodrich 2012). Nitidulid beetles are active during the day and crawl into the tightly enclosed floral chambers during the first day of anthesis, making contact with the numerous receptive stigmas. Beetles usually leave the chamber on the second day when the protogynous flowers are in a pollen-shedding phase and scent emission declines. The floral biologist Gerhard Gottsberger, who has made a lifelong study of beetle pollination in the tropics, has suggested that fruit mimicry may be the ancestral condition in the Annonaceae, based on evidence that early divergent genera (those in lineages that

Figure 6.12 Flowers of *Duguetia stelechantha* (Annonaceae) smell like fermented fruit and are pollinated by *Colopterus* nitidulid beetles (an individual is indicated by the arrow) in its native range in South America. Beetles enter the closed floral chamber formed by the internal petals. Photo © Gerhard Gottsberger.

split from the main family lineage early on), such as *Anaxagorea*, are pollinated by nitidulids.

Small-flowered Annonaceae pollinated by nitidulids are seldom thermogenic, in contrast to the much larger-flowered thermogenic species pollinated at night by dynastine scarab beetles. These scarab beetles are seemingly attracted to a different set of volatiles and often feed on fleshy flower parts. The motivation of dynastine beetles for visiting these specialized flowers is not well understood. In particular, it is not clear whether dynastine scarab beetles are searching for oviposition sites when they visit these flowers or whether they depend on the flowers simply as a food resource.

Cones of the South African cycad *Stangeria eriopus* emit a scent reminiscent of rotting bananas and are also pollinated by nitidulid beetles (Proches and Johnson 2009). The beetles crawl within the cones of both sexes and effectively transfer pollen. Drosophilid flies are also attracted, but are too large to enter the narrow spaces between the cone scales. Analysis of the scent showed that it is dominated by acetic acid esters. Pollination by nitidulid beetles has also been reported in the Asian cycad *Cycas revoluta* (Kono and Tobe 2007). The main scent compound in *C. revoluta* was reported as being the phenylpropanoid estragole (Azuma and Kono 2006), but a later analysis suggested that it is dominated by

the fermentation volatile ethyl acetate (Proches and Johnson 2009). The ester isoamyl acetate imparts the distinctive banana odor of both *S. eriopus* and *C. revoluta*. Unlike other cycads with nursery pollination systems, these cycad species are largely deceptive. Their pollination systems, involving mimicry of the angiosperm fruit brood sites of nitidulid beetles, are a unique instance of mimicry among gymnosperms.

Interestingly, unlike most other cycads, *Stangeria* does not have thermogenic cones. This provides an interesting parallel with the Annonaceae in which the nitidulid-pollinated flowers are generally not thermogenic, while those pollinated by dynastine scarabs are thermogenic (Gottsberger 2012).

The existence of pollination by sap beetles in basal angiosperms and some cycads can be taken as an indication of an ancient origin of pollination by these insects. The Nitidulidae as a group probably originated in the late Mesozoic (Grimaldi and Engel 2005) and sap beetle pollination in *Stangeria* and *Cycas* may be a modern convergence with similar pollination systems in angiosperms. However, beetles would have used rotting plant material for brood sites long before the evolution of fleshy fruits and sap beetles. It is therefore possible that floral mimicry of the scent of rotting vegetable matter (and the volatiles produced by bacteria and yeasts feeding on such material) followed the same evolutionary trajectory as floral mimicry of dung oviposition sites, i.e., as a response to a pre-existing and ancient bias in the sensory systems of beetles (Schiestl and Dötterl 2012).

Fruit flies belonging to the family Drosophilidae are a much younger group than most of the beetle lineages that feed on fruits. They too are exploited by flowers which mimic the signals of fermenting fruit. A well-studied example is the Mediterranean Solomon's lily, *Arum palaestinum*, which has a pleasant odor reminiscent of "fruity wine," quite unlike related *Arum* species that smell like dung. In the few hours during which it produces its odor, a single plant of *A. palaestinum* can attract hundreds of drosophilid flies. The taxonomic composition of this assemblage of at least eight drosophilid species is very similar to that attracted to rotting bananas. The volatiles produced by *A. palaestinum* are in fact very like those produced by yeasts. Six compounds—2,3-butanediol acetate, acetoin acetate, ethyl hexanoate, hexyl acetate, 2-phenylethyl alcohol, and 2-phenylethyl acetate—produced highly reproducible physiological responses in the antennae of *Drosophila melanogaster*, and a mixture of these compounds was as attractive to the flies as banana bait (Stökl et al. 2010). None of the compounds by themselves was as effective as the mixture. 2,3-Butanediol acetate and acetoin acetate are particularly interesting components of the scent because they could specifically signal the presence of yeasts (the compounds are derived by acetylation of the yeast-derived compounds 2,3-butanediol and acetoin). Calcium imaging of the brains of flies showed that both highly conserved odorant receptors (relating to fermentation products) and divergent odorant receptors (relating to various fruit habitats) are involved in the flies' responses (Stökl et al. 2010). This explains why flies find the signal of the Solomon's lily irresistible and also how such a broad assemblage of flies can be attracted.

A much more specialized oviposition-site mimicry system exploiting drosophilid flies has been described in the orchid *Gastrodia similis* (Martos et al. 2015). Rather than targeting a broad range of drosophilids, this orchid relies on a single fruit fly species (*Scaptodrosophila bangi*) for pollination, and has a scent dominated by just three esters—ethyl acetate, ethyl isobutyrate, and methyl isobutyrate. Specialization in this system is also a consequence of a very intricate floral mechanism that functions only with flies of a particular size. Flies landing on the labellum of the orchid are captured by a touch-sensitive trapdoor mechanism that briefly imprisons them in a space between the labellum and the column (Martos et al. 2015). The narrow exit from the chamber precisely matches the width of *S. bangi* and brings the dorsal surface of the flies into contact with the pollen transfer apparatus of the flower. This trap therefore serves not only to sort insects according to the correct size, but also to manipulate them into the correct position for pollen transfer. However, it is not the trapping device that is the primary basis of the specialization of the orchid for a single pollinator species. Bioassays in which synthetic mixtures of the three esters emitted by the flowers were placed alone and in combination were highly attractive to *S. bangi* and its congener *Scaptodrosophila triangulifer*. Two of these esters were identified in fermenting *Ficus* fruits that attract *Scaptodrosophila* flies. Other drosophilids, including *Drosophila* species, were attracted to banana baits yet were not found on the orchid or in the bioassay experiments, thus suggesting that the orchid selects *Scaptodrosophila* species from the local assemblage of fruit flies.

There is now good evidence that the attraction of many *Drosophila* species to fermenting fruits is mediated by the volatiles emitted by yeasts rather than by the fruit itself (Becher et al. 2012). Indeed, yeasts benefit—in terms of colonization opportunities and outbreeding—from directed dispersal by *Drosophila* to fruits. It has even been suggested that yeasts produce volatiles not only as a simple by-product of fermentation but also as a signal advertising potential brood sites, in order to maximize their own dispersal by *Drosophila* species (Arguello et al. 2013; Christiaens et al. 2014). Deletion of the gene controlling the production of acetate esters (specifically ethyl acetate and isoamyl acetate) dramatically reduced the dispersal of *Saccharomyces cerevisiae* yeasts by fruit flies (Christiaens et al. 2014).

Experiments conducted with *Drosophila melanogaster* have shown that yeasts are responsible for volatile signals that attract flying drosophilids from a distance, and also increase both egg-laying by adult flies and the successful development of fly larvae (Becher et al. 2012) (Fig. 6.13). It was found that a mixture of five volatiles—ethanol, acetic acid, acetoin, 2-phenyl ethanol, and 3-methyl-1-butanol—was as attractive to *D. melanogaster* as fermenting grape juice, and thus likely to constitute the core signal used by this fly to locate its oviposition sites. There was a small degree of attraction to yeast-free substrates, particularly unfermented grape juice, which helps to explain how yeasts may be originally vectored to fruits. One of the unanswered questions is whether yeast-free substrates may attract fruit flies through volatiles produced by bacteria that had initially colonized the substrate. Another question is whether fruit flies that are more specialized in their choice of fruits for oviposition might use other cues such as fruit esters that are not produced by yeasts.

(a)

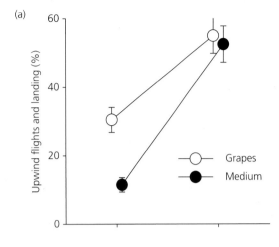

(b)

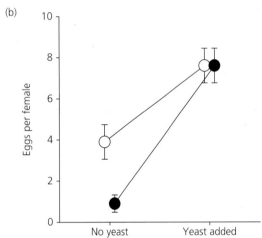

Figure 6.13 The influence of inoculation of substrates with yeast on attraction and egg-laying of the fruit fly *Drosophila melanogaster*. Addition of yeast to grapes or a growth medium markedly increases (a) the number of upwind flights and landing on a source in a wind tunnel and (b) the number of eggs laid by female flies. Adapted from Becher et al. (2012).

Mushroom mimicry

The use of fungi as a brood site for larval development, particularly the fruiting bodies of basidiomycete mushrooms, has evolved in a number of fly lineages including the Sciaridae and Mycetophilidae ("fungus gnats") and the Drosophilidae. The cues used by these mycophagous insects to locate oviposition sites are deployed by some flowers as part of a strategy of fungal mimicry. This

extraordinary form of floral deceit was discovered by Stefan Vogel in the course of his ground-breaking work on the pollination of *Asarum caudatum* (Plate 8), a pipevine native to the moist forests of western North America (Vogel 1973, 1978). He confirmed that the dark-colored flowers that are produced on the forest floor are pollinated by mycetophilid fungus gnats. The gnats oviposit within the flowers but the emerging larvae soon perish. Vogel speculated that oviposition is elicited by the combination of olfactory, visual, tactile, and humidity cues, the latter arising from high rates of transpiration within the flowers. Pollination by mycetophilid fungus gnats was later shown to occur in the related Asian pipevine *Asarum tamaense* (= *Heterotropa tamaensis*) (Sugawara 1988). Fungus gnat eggs were frequently found in the flowers, confirming that pollination occurs through mimicry of an oviposition substrate.

Vogel (1973, 1978) demonstrated that fungal mimicry has evolved independently in aroids, such as the Indian Sikkim cobra lily, *Arisaema utile*, and the Mediterranean species *Arisarum proboscideum*, also known as the mouse-tail plant. Several *Arisaema* and *Arisarum* species have a clavate spongy white appendix that is strikingly fungus-like in appearance, and Vogel (1973) confirmed that mycetophilid flies lay their eggs in the depressions of the "fungoid" appendix of *Arisarum proboscideum*.

The aroid genus *Arisaema* contains approximately 150 species and is centered in the Himalayas and southwestern China. Vogel and Martens (2000) showed that pollination by mycetophilid and sciarid flies is almost ubiquitous in *Arisaema*. Most species are functionally dioecious and have a phenotypically plastic form of sex expression, with young plants being male and older plants female (Bierzychudek 1982). Flies that slide on the waxy surface of the spathe and fall into the chamber eventually escape from an exit hole in the spathe, but female spathes provide no such exit (indeed, there is no selective benefit for female spathes to allow flies to escape) and flies remain thus detained until they die. *Arisaema* species are not very selective in terms of the fungus gnats that they attract, and plants cultivated in Europe attract an assemblage of gnats different from that in their native range in Asia (Vogel and Martens 2000).

Pipevines and aroids that mimic fungal oviposition substrates exhibit striking convergence in a number of features in addition to their shared possession of floral chambers. One such feature is the floral coloration, which is frequently green or brown with a pattern of white stripes. Another is the flagelliform apices of the perianth segments that are up to 80 cm long in some species of *Arisaema* (Vogel and Martens 2000). These are often scented and make contact with the substrate. Vogel and Martens argued that they serve as "conducting paths" along which the gnats travel.

The most compelling evidence for chemical mimicry of the fruiting bodies of mushrooms has come from studies of the large Neotropical orchid genus *Dracula*, originally spearheaded by the Swiss natural product chemist Roman Kaiser (2006) and, more recently, by a team of fungal ecologists (Policha et al. 2016). The scent of *Dracula* flowers is emitted mainly from the labellum and is typically characterized by eight-carbon alcohols and ketones (e.g., oct-1-en-3-ol, oct-1-en-3-one, octan-3-ol, and octan-3-one) that are typical of the odor of mushroom fruiting bodies (Kaiser 2006; Policha et al. 2016). Mycophilous insects are known to use these compounds to locate fungal fruiting bodies (Faldt et al. 1999). Not only do *Dracula* flowers smell like mushrooms, but the labellum of many species bears a striking visual resemblance to mushrooms, including the white color and crinkled appearance of the gills of mushroom caps. These transverse ridges are very similar to those found within the flowers of *Asarum* species that mimic fungi (Vogel 1973, 1978; Sinn et al. 2015).

Dracula orchids are pollinated by drosophilid flies in the genera *Zygothrica* and *Hirtodrosophila* (Endara et al. 2010). These flies are closely associated with mushrooms, but the life histories of most species are not yet known. In their work on *Dracula lafleurii*, Policha et al. (2016) recorded at least 11 drosophilid fly species that remove pollinaria from the flowers; many of these species were also found on mushrooms in the same habitat. They surveyed the visual and scent properties of a broad range of mushroom species and concluded that *Dracula* species employ a generalized system of mushroom mimicry, a strategy that makes sense given the ephemeral and unpredictable appearance of mushroom fruiting bodies. To establish the role of visual and olfactory cues in this system, they constructed precise silicone replicas of the flowers using 3D-printing technology and then altered the visual and scent signals of these model flowers. Model flowers with extracts of the scent of *Dracula* flowers were more attractive to the flies than those lacking this scent (Fig. 6.14). Flies also preferred spotted model flowers over models colored plain green, white, or red (Fig. 6.15). These findings suggest that both visual and olfactory signals are important for effective floral mimicry of mushrooms. What remains unknown is whether the mushroom-like shape of the labellum is important for attracting flies.

The behavior of flies, including courtship by wing movements (semaphoring), mating, and proboscis extension, was essentially the same on mushrooms and *Dracula* flowers. In addition to using the mushroom fruiting bodies as a brood site, mycophilous drosophilids also feed on yeasts growing on their surface (Policha 2014). Thus the mimicry of mushrooms by *Dracula* flowers may combine elements of brood-site and food-source mimicry, in much the same manner as flowers that mimic carrion or dung.

Aphid mimicry

Adult hoverflies are well-known flower visitors that feed on both pollen and nectar. Some hoverflies have larvae that feed on aphids (and other homopterans); adult females of these aphidophagous species seek out aphid colonies on which to lay their eggs. There is now evidence that the flowers of some plant species mimic the visual and chemical properties of aphid colonies in order to exploit aphidophagous hoverflies as pollinators. This novel form of oviposition-site mimicry is so far known only from a number of orchid lineages.

The first suggestion of aphid mimicry by flowers was made by Ivri and Dafni (1977) in their study of the pollination of the orchid *Epipactis veratrifolia* (= *E. consimilis*) in Israel. They noted that hoverflies frequently laid eggs alongside the dark aphid-like swellings on the labellum, even though aphids were seldom seen on flowers. Very small amounts of nectar are produced and pollination occurs when hoverflies lick at this nectar. Male hoverflies

(a)

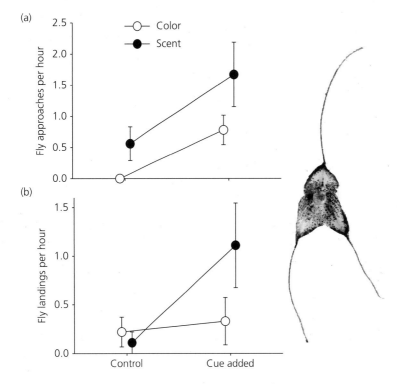

(b)

Figure 6.14 The combined effects of visual and olfactory cues on the attraction of fungus-associated drosophilid flies to artificial silicone flowers (insert) in a rainforest in Ecuador. The artificial flowers created using 3D-printing technology matched the morphology of flowers of the orchid *Dracula lafleurii*, which is a mimic of mushrooms. The control for artificial flowers painted to resemble the orchid was plain green. The scent cue added was a solvent extract of the orchid flowers. (a) Number of fly approaches per hour. (b) Number of fly landings. Photo © Melinda Barnadas. Adapted from Policha et al. (2016).

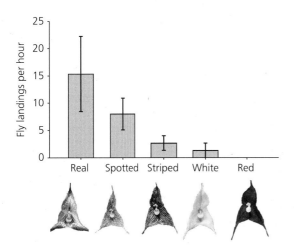

Figure 6.15 The effects of different visual cues on attraction of fungus-associated drosophilid flies to artificial flowers matching the morphology of *Dracula lafleurii*. Unscented artificial flowers with maroon spots on a white background came close to being as attractive to the flies as real flowers, and were significantly more attractive than those with maroon stripes on a white background or those that were plain white or maroon red. Only the central portion of the artificial flowers is shown in the figure. Photos © Barbara Roy. Adapted from Policha et al. (2016).

establish territories around the flowers where they try to copulate with approaching females and they also pollinate flowers while feeding on nectar.

Female aphidophagous hoverflies use both visual and olfactory cues to locate aphids, but egg-laying itself requires olfactory cues (Dixon 1959). The flowers of *E. veratrifolia* emit a blend of terpenoid compounds which is similar to that of the alarm pheromones of several aphid species (Stökl et al. 2011). Electrophysiological experiments show that these terpenoids are detected by the antennae of hoverflies of both sexes. Headspace samples of the scent of the orchid flowers increased egg-laying by hoverflies, as did a synthetic mixture of four terpenoid compounds that are produced by the flowers (Stökl et al. 2011) (Fig. 6.16). Since the chemical composition of the scent of *E. veratrifolia* is not identical to that of any particular aphid alarm pheromone, the interpretation is that the orchid has a generalized system of mimicry of aphids, akin to the generalized mimicry of feces, carrion, or mushrooms by other plants with oviposition-site mimicry systems.

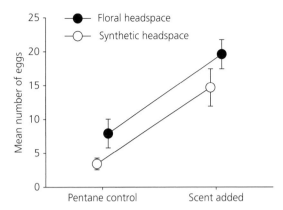

Figure 6.16 The effect of volatiles on oviposition behavior of female hoverflies (*Episyrphus balteatus*) on bean plants (*Vicia faba*). Addition of headspace extracts from flowers of the hoverfly-pollinated orchid *Epipactis veratrifolia* or a synthetic mixture of terpenoids recorded from the floral headspace led to similar increases in the number of eggs laid. Adapted from Stökl et al. (2011).

Aphids feed on the sap of vegetative parts of *Epipactis* plants, and one intriguing possibility is that the emission of terpenoids by flowers evolved originally as means of defending the reproductive parts of the flowers against aphids, which are deterred by these compounds. The system may then have evolved into an oviposition-site mimicry system as hoverflies responded positively to these compounds and increased the pollination success of plants. Flowers of other forms of *E. veratrifolia* in the eastern Himalayas appear to be regularly infested with aphids, and it has even been suggested that hoverflies may sometimes benefit from laying eggs on this form of the orchid (Jin et al. 2014).

Mimicry of aphid brood sites appears to have evolved independently in some Asian and South American slipper orchids. In a study conducted in Borneo, Atwood (1985) found up to 76 hoverfly eggs on the staminode of *Paphiopedilum rothschildianum* (Plate 8d). He suggested that the densely hairy staminode may mimic an aphid colony. However, Bänziger (2002) and Bänziger et al. (2012) pointed out that the brown dots, and especially the hairy wart-like structures, on the petals of many *Paphiopedilum* species were far more reminiscent of aphids. However, Bänziger and his co-workers found no evidence that flies were attracted directly to these structures. They pointed out that a scent cue

is probably responsible for attraction of hoverflies and that this likely originates from the trichomes on the staminode. Hoverflies fall into the labellum chamber of *P. rothschildianum* when attempting to fly from the slightly glutinous staminode and pick up or deposit pollen when passing through the narrow exit. Atwood did not find aphids on plants of *P. rothschildianum*, and no nectar is produced by them. The system is therefore entirely deceptive. Bänziger (2002) reported similar findings from his study of *Paphiopedilum callosum* (Plate 8c) in Thailand. No aphids were found on any of the orchids examined, yet up to 40 eggs were found on some flowers and these were traced to visits by females of the aphidophagous hoverfly *Episyrphus alternans* (Fig. 6.17).

The role of brood-site mimicry in some other *Paphiopedilum* species pollinated by hoverflies is less clear-cut. *Paphiopedilum villosum*, for example, is pollinated mainly by females with aphidophagous larvae, yet no eggs were found to be laid on the flowers and some non-aphidophagous hoverflies were trapped in the flowers (Bänziger 1996a). The hoverflies are attracted to a glistening patch on the bright yellow staminode, in the middle of which is a knob-like protuberance (the "mesmerizing wart") on which hoverflies attempt to land. As it is slippery, the flies lose their grip and fall into the labellum chamber. This glistening patch may resemble aphid honeydew (known to be used as food by some hoverflies) or nectar. Bänziger et al. (2012) have shown that some related *Paphiopedilum* species—*P. bellatulum*, *P. concolor*, and *P. godefroyae*—are pollinated by herbivorous milesiine hoverflies, rather than predatory syrphine hoverflies. He cautioned that black spots should not automatically be assumed to mimic aphids, particularly since the majority of aphids are green and yellow and have little contrast against foliage, yet are readily discovered by hoverflies. *Paphiopedilum dianthum* has light-colored patterns on the staminode as well as darker spots on the petals and is pollinated by gravid females of the hoverfly *Episyrphus balteatus* (Shi et al. 2007). These hoverflies lay eggs mostly near the papillate tips of the petals, the likely sources of odor, and not close to the dark spots on the petals. Interestingly, this same hoverfly species was also implicated as a pollinator of the congener *Paphiopedilum barbigerum*. This species has a yellow staminode matching the

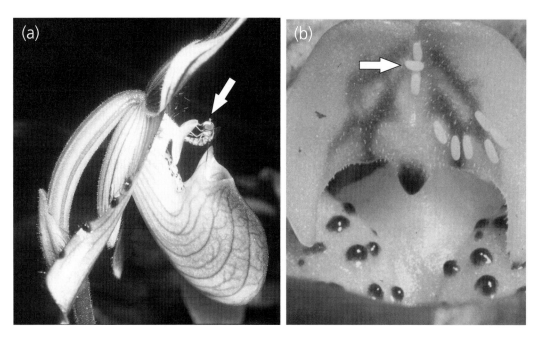

Figure 6.17 The Asian slipper orchid *Paphiopedilum callosum* attracts aphidophagous hoverflies through mimicry of the signals produced by aphids. (a) A female *Episyrphus alternans* hoverfly (arrowed) attempts to lay an egg moments before tumbling into the lip chamber. (b) Hoverfly eggs (arrowed) on the staminode of *P. callosum.* Adapted from Bänziger (2002). Photos ©Hans Bänziger.

visual properties of anthers of flowers used as pollen sources by the hoverflies, and is thus likely mimicking pollen (Shi et al. 2009). However, the overwhelming majority of hoverflies responsible for pollinating *P. barbigerum* are female, raising questions about whether brood-site cues may also play a role in those *Paphiopedilum* species pollinated by hoverflies that do not lay eggs (Bänziger et al. 2012).

Mimicry of aphids has also been suggested to occur in the South American slipper orchid genus *Phragmipedium*. Species with dull-colored (green and brown) flowers often have aphid-like markings on the folded lip margin, and Pemberton (2011) has observed female hoverflies flying against the markings on flowers of *Phragmipedium pearcei* and ricocheting into the labellum pouch. The flies then acquire a pollen smear as they squeeze themselves out of the flower (Plate 8e).

Oviposition-site mimicry in fungi and mosses

Flies that use feces and carrion as sources of food and as sites for oviposition are exploited not only by angiosperm flowers but also by fungi and mosses (Marino et al. 2009; Johnson and Jürgens 2010). The function of fly attraction in fungi and mosses is to disperse spores. In fungi this is achieved when flies consume an exudate (the gleba) that contains the spores, which germinate once they have passed through the flies' digestive systems. In the case of mosses the spores are sticky and are usually transported on the bodies of the flies.

Stinkhorn fungi belonging to the family Phallaceae have notoriously foul-smelling fruiting bodies, and chemical analysis has shown that they emit the typical components of the smell of feces and dead animals (Borg-Karlson et al. 1994; Johnson and Jürgens 2010). Mosses of many species in the family Splachnaceae have scented sporophytes that are visited by flies (Marino et al. 2009). These mosses often grow in decaying substrates such as carrion or feces; thus mimicry of the odor of feces or dung can be viewed as part of a strategy that ensures dispersal to suitable germination sites for the spores. Unlike the case in stinkhorns which produce an edible gleba, the mosses offer no rewards to flies. Moss species vary in the chemical profile of the volatiles emitted from the sporophytes; some mainly attract flies associated with carrion, such as Sarcophagidae and Calliphoridae, while others primarily attract flies associated with dung, such as Muscidae and Scathophagidae (Marino et al. 2009). Other species have bright-colored parasol-like sporophytes with weaker odors and attract a general fly fauna, including many anthomyiid flies (Marino et al. 2009).

Oviposition-site mimicry and diversification of plant lineages

The evolutionary radiation of several major plant lineages with a combined total of thousands of species has been closely associated with the evolution of oviposition-site mimicry. These groups include the custard apples (Annonaceae), pipevines (Aristolochiaceae), aroids (Araceae), stapeliads, ceropegias, and brachystelmas (Apocynaceae: Ceropegieae: Stapeliinae), bulbophyllums (Orchidaceae: Bulbophyllinae), pleurothalids (Pleurothallidinae), and Rafflesiaceae. A major driver of diversification in these groups has undoubtedly been the huge numbers of insect oviposition niches available for plants to exploit.

In addition to shifts between different oviposition-site mimicry systems, diversification in many of the aforementioned groups has been promoted by shifts between deceptive and mutualistic pollination systems. Examples include shifts between fruit mimicry involving sap beetles and rewarding systems involving dynastine scarabs in Annonaceae (Gottsberger 2012), between deception of flies and brood-site mutualisms in pipevines (Sakai 2002), and, in aroids, between several very different systems—deception of flies and beetles seeking carrion or dung, pollen rewarding of bees, and brood-site mutualisms involving drosophilids (Diaz and Kite 2006).

In *Asarum*, a genus belonging to the pipevine family (Aristolochiaceae) and containing 100 species, diversification is correlated with a reduction in vegetative growth, loss of autonomous self-pollination and the evolution of fungal-mimicking structures (Sinn et al. 2015).

In the Araceae, undoubtedly the best-studied plant group in terms of oviposition-site mimicry, there were at least five origins of deceptive pollination systems, two associated mostly with beetles (in *Stylochaeton* and *Amorphophallus*) and three mostly with flies (*Arisarum* and the Areae and Cryptocoryneae clades) (Chartier et al. 2014). Within each of these clades, shifts occurred between different types of oviposition-site mimicry (Stökl et al. 2010). The evolution of unisexual flowers in the Araceae led to separate zones of male and female florets along the spadix and the repeated evolution of a trap by a central restriction in the spathe (Bröderbauer et al. 2012). These traps allow plants to achieve accurate pollen delivery by large numbers of small insects and largely negate the need for floral rewards. Another important innovation that facilitated deception was the sterile appendix, which serves both as a site for thermogenesis and scent release and as a landing and take-off place for insects.

Other lineages such as the Rafflesiaceae appear to have undergone shifts between various modes of carrion mimicry (Bänziger 1996b; Davis et al. 2007), but have not shifted to other modes of pollination. Is the floral trait space that is occupied by oviposition-site mimicry so unusual that these flowers are unlikely to be visited by other pollinator groups, and thus do not allow for the stage of double-function that is required for most pollinator shifts (Stebbins 1970)? Is it possible, therefore, that oviposition-site mimicry is often an evolutionary dead-end, despite its repeated evolution across many lineages?

Overview and perspectives

Oviposition-site mimicry is undoubtedly a major driver of diversification in floral volatile signals (Jürgens et al. 2006; Ollerton and Raguso 2006; Jürgens et al. 2013). There is now ample evidence that the success of this form of deception also depends on the evolution of sophisticated morphological and color signaling traits (Policha et al. 2016). Indeed, some of the most morphologically derived flowers in the world are brood-site mimics (Vogel 1978; Barkman et al. 2008).

Despite all the progress that has been made in deciphering the chemical and visual codes underlying brood-site mimicry, there are still some major unanswered questions. Perhaps the most important of these is whether brood-site mimicry is clearly separable from food-source mimicry. While the presence of eggs and larvae of insects on some flowers provides a clear indication that insects have cognitively misclassified the flower as a brood site, it is not always possible to extrapolate this conclusion to other insects that visit the flower or to other related systems. It is well known that many of the substrates used by insects as oviposition sites are also used as food sources, often by the same insects that use them for oviposition (Bänziger and Pape 2004).

A good case in point is the uncertainty surrounding the characterization of floral mimicry of animal dung. Some of the problem areas include: (1) the lack of absolute chemical differences in the scent profiles of carrion and dung (Jürgens et al. 2013), (2) the shared use of both carrion and dung for oviposition and larval development by some fly and beetle species (Bänziger and Pape 2004), and (3) the use of fresh feces as a food source by insects, including those that oviposit exclusively on carrion (Brodie et al. 2016). The latter phenomenon means that a flower that smells like feces may therefore attract insects seeking oviposition sites, as well as those seeking food. Of course, this is a problem only for biologists seeking a neat system of classification; a plant with flowers that smell like feces will usually benefit in terms of fitness if it attracts both gravid and food-seeking insects.

We have also seen that there are likely to be biochemical trade-offs relating to the mimicry of different substrates. For example, gravid blow flies seeking a specific oviposition site on carrion may be repelled by a flower that smells like both carrion and feces (Fig. 6.10). We feel that some of the most exciting future research in the field of floral mimicry will be focused on developing an understanding of the ultimate and proximate bases of specialization in oviposition-site mimicry systems.

There are also many unanswered questions about the biochemical basis of mimicry in oviposition-site mimics. How do plants that mimic the oviposition sites of sap beetles and fruit flies produce typical yeast volatiles such as ethanol, acetoin, isoamyl acetate, and ethyl acetate? Do they encourage colonization of the flowers by yeasts and provide them with sugars so that these compounds are emitted, or do they produce these compounds from their own biochemical machinery (Goodrich et al. 2006)? Similar questions can be raised about carrion mimicry, as the volatile cues in these systems are usually the products of bacterial activity (Schulz and Dickschat 2007).

The functional significance of the mottled pattern that has evolved convergently in many different oviposition-site mimics (Fig. 6.9) also remains a mystery. Is it a signal, as suggested by some experiments (Fig. 6.15), or is it a form of crypsis that renders flowers less likely to be discovered by non-pollinators?

Perhaps the greatest limitation to our understanding of oviposition-site mimicry relates to our lack of knowledge about the life histories of insects, particularly tropical species. Without this knowledge, it becomes very difficult to identify putative model organisms and therefore to use a niche-based approach to the evolutionary diversification of the mimics. To characterize these niches, we also need a better understanding of the ecological succession of microbial communities on various substrates, as these microbes produce many of the volatile cues that are imitated by flowers.

Combine these questions with some of the others raised previously in this chapter, such as the uncertainties around the function of thermogenesis, the adaptive significance of floral gigantism, and the role of tactile and visual cues in carrion mimicry, and it becomes obvious that we are still far from having a complete understanding of oviposition-site mimicry in plants.

CHAPTER 7

Special cases

Biology is the science of such exceptions in a way that physics is not

Forbes (2009, p. 56)

Introduction

We started this book by outlining the general principles of floral mimicry and then attempting to unify these principles with those developed for protective mimicry in animals. Most of the subsequent chapters dealt with floral mimicry systems that are, to varying degrees, conceptually aligned with Batesian mimicry. This chapter deals with topics that are contentious, difficult to categorize, or fall somewhat outside the mainstream of floral mimicry research.

The first topic is intraspecific resemblance between flowers of different sexes. How do we identify similarity that goes beyond mere shared ancestry and common genetic architecture? In cases where only one sex offers a reward, the nonrewarding sex which is visited by "mistake" (Baker 1976) is usually under strong selection to resemble the rewarding sex, thus constraining the evolution of dimorphism. Is this an example of mimicry?

The second topic concerns resemblance among rewarding flowers of unrelated species in the same community. Such resemblance is primarily related to adaptations to common pollinators and falls under the umbrella of floral syndromes. However, there is an intriguing possibility that selection for resemblance may also play a role in the evolution of the phenotypes of rewarding species. We survey the available evidence and identify the types of tests that would be required to resolve this conundrum.

The third topic relates to the question of whether fungal pseudoflowers (flower-like structures produced for sexual reproduction or dispersal of spores) should be considered to be an example of mimicry. This topic strictly lies outside the field of mimicry in plants since these flower-like structures are produced by a separate kingdom, much like the flower-like structures produced by certain mantids that exploit the sensory bias of flower-visiting insects (O'Hanlon 2014; O'Hanlon et al. 2015). However, the fungi that produce these pseudoflowers are parasites of flowering plants and attract flower-visiting insects for purposes of reproduction (not predation as is the case in mantids), and we felt that this was sufficiently relevant to warrant its inclusion in this book.

We also deal with two topics that did not fit neatly into any of the other chapters. The first of these concerns flowers that mimic the chemical signature of leaves under attack by insect herbivores and thereby attract wasps searching for caterpillars to feed to their offspring. This is a system that combines elements of food-source and oviposition-site mimicry. The last topic concerns flowers that are pollinated mainly by male insects yet do not appear to be sexually deceptive.

All of the topics included in this chapter require further research. We therefore summarize what is known about these special cases in order to develop a platform of knowledge for the future.

Floral Mimicry. Steven D. Johnson & Florian P. Schiestl.
© Steven D. Johnson & Florian P. Schiestl 2016. Published 2016 by Oxford University Press.
DOI 10.1093/acprof:oso/9780198732693.001.0001

Intersexual resemblance

Sexual dimorphism is characteristic of many animals and plants and has been the subject of much research, particularly that focused on the role of sexual selection. Intersexual resemblance, on the other hand, seems far less remarkable in flowering plants than it does in animals, mainly because it could be considered the default condition for two sexes that originate from a hermaphrodite ancestor.

In the case of animals, which are often strongly dimorphic, resemblance between the sexes is usually interpreted as a derived condition. Cases of intraspecific sexual resemblance in animals that have been interpreted as mimicry mostly involve males that mimic females in order to penetrate the territories of other non-mimetic males and thereby gain mating opportunities. Some male scorpion flies even mimic females in order to rob males of their nuptial gifts (Thornhill 1979). In many damselfly species, a proportion of the females mimic males; this may afford such females an advantage in terms of escaping harassment by males (Gosden and Svensson 2009).

In dioecious and monoecious plants, sexual selection in the form of male–male competition for access to ovules can lead to exaggerated male floral traits and thus gender dimorphism (Willson 1979; Bawa 1980a; Romero and Nelson 1986). This is consistent with theory that predicts that male fitness should be limited mainly by mating opportunities while in females fitness should be limited mainly by resources for provisioning seeds. The trend for male flowers to be particularly showy applies not only to visual signaling traits but also to scent (Ashman 2009). In the dioecious species *Silene latifolia*, for example, male flowers produce more scent than female flowers and are more attractive to pollinators (Waelti et al. 2009). One of the simple indicators for strong male–male competition in dioecious plants is that males tend not only to have flowers that are more showy and have greater rewards than those of females but also to produce a larger number of flowers (Bawa 1980b; Bond and Maze 1999).

Dimorphism in animal-pollinated plants with unisexual flowers should generally be constrained because both sexes need to be attractive to the same flower visitors (this constraint does not apply in animals since they mate directly). Differences in signals between sexes that result in pollinators showing constancy to one sex is likely to reduce the efficiency of pollen transfer. Kay (1982) showed that flower-visiting insects often show preference for one sex when foraging on plants with unisexual flowers. Observations of dioecious *Silene dioica*, for example, showed that pollen-collecting honey bees prefer male flowers, while nectar-feeding *Bombus* species prefer female flowers. This problem of sex-specific foraging constancy is obviously greatly exacerbated when only one sex offers a reward (Ågren and Schemske 1991). A common situation in plants with unisexual flowers is for males to offer a pollen reward and for females to be pollinated by deceit (Willson and Ågren 1989; Renner and Feil 1993). Herbert Baker (1984) even suggested that in lineages where nectar production has been lost, the evolution of unisexual flowers would be strongly constrained because of sex-specific foraging by pollinators. In the genus *Solanum*, which includes several instances of the evolution of dioecy in a pollen-rewarding lineage, female flowers produce non-functional pollen grains that reward flower visitors (Anderson 1979).

In addition to the typical situation where males produce pollen and females rely on deceit, there are also many examples of dioecious and monoecious plant species in which nectar is produced by only one sex or in which only one sex offers a brood site for pollinators (Willson and Ågren 1989). In all such cases we would expect selection in the non-rewarding sex to favor similarity to the rewarding sex. Pasteur (1982) referred to adaptive interspecific resemblance among unisexual flowers as "Bakerian mimicry" in honor of Herbert Baker's (1976) work on "mistake pollination" of non-rewarding female flowers in the Caricaceae.

Similarity in the floral displays of the male and female flowers of dioecious and monoecious species is usually due to shared homologous features. This in itself is not an indication of mimicry. The strongest arguments for intersexual mimicry are based on modification of the structures of one sex to resemble non-homologous structures of the other sex (Bawa 1980b; Dukas 1987; Willson and Ågren 1989; Grison-Pige et al. 2001; Soler et al. 2012). For example, in *Begonia involucrata*, pollen-collecting

bees may visit the rewardless female flowers because they have a stigma that is modified to resemble pollen-bearing anthers (Ågren and Schemske
1991). In the monoecious tree *Jacaratia dolichaula*
(Caricaceae), female flowers are non-rewarding and
lack a corolla tube, but have stigma lobes that are
modified to resemble the white corolla tubes of the
rewarding male flowers (Bawa 1980b). Of special interest to us here is whether intersexual resemblance
in floral signals is maintained by selection because
pollinators are more likely to move between the
sexes and thus transfer pollen. Although the theory
of sexual selection is based on the premise that
females are limited mainly by resources, there is
now strong evidence that female fitness in plants
is often limited by mating opportunities (Knight
et al. 2005). This phenomenon of pollen limitation
helps to explain why female unisexual flowers often have elaborate adaptations that increase their
similarity to males.

Schemske and Ågren (1995) tested whether bees
prefer female *Begonia* flowers that most closely resemble rewarding male flowers in terms of size
(mimicry hypothesis) or show directional selection
for increased flower size (Fig. 7.1). Their results for
Begonia involucrata, based on choices among artificial flowers, showed clearly that bees were more
likely to visit larger female flowers. Importantly,
flower size tended to decrease with flower number,
suggesting that the upper limit to flower size might
be constrained by resource availability. A very similar result of bee preference for larger female flowers was also obtained for another species, *Begonia
oaxacana* (Schemske et al. 1996). In this system bees
show a strong overall preference for male flowers (they visit five male flowers for every female
flower), which suggests that the deception is not
complete. The likely explanation for the preference
for larger female flowers in these two *Begonia* species is that bees are conditioned by positive associations between flower size and rewards in male
flowers (Schemske et al. 1996). Another factor may
be the greater conspicuousness of larger flowers to
bees. Somewhat different results were obtained by
Castillo et al. (2012), who found that manipulated
female *Begonia gracilis* flowers that most closely re-
sembled the most attractive male phenotype had
the highest probability of setting fruit. They showed

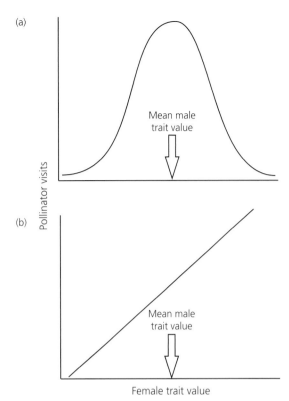

Figure 7.1 Alternative hypotheses for selection on signaling traits
of non-rewarding female flowers. (a) Stabilizing selection when
pollinators prefer female flowers with trait values closest to the
mean value of rewarding male flowers. (b) Directional selection when
pollinators prefer female flowers with larger trait values, regardless of
the male mean. Pollinator preference for larger trait values may arise
when there are correlations between trait values and rewards in male
flowers. Adapted from Schemske and Ågren (1995).

that perianth size alone is not an accurate indicator
of androecium size in this species and that pollinators preferred male flowers where the androecium
was large relative to the perianth. This foraging
strategy by bees may explain why female flowers
with a small perianth and relatively large stigmatic
lobes were most successful in terms of setting fruit.

The degree of discrimination among flowers of
different sexes also depends on the memory and
discrimination ability of pollinators. Dukas (1987),
for example, found that solitary bees were far more
likely than honey bees to mistakenly visit the rewardless female flowers of the cucurbit *Ecballium
elaterium*. He found that honey bees were more

likely to visit female flowers in the afternoon, and suggested that the decrease in visible pollen in male flowers made it harder for honey bees to discriminate between the sexes at that time of day. Similarly, Ågren et al. (1986) found differences among insect groups in terms of their discrimination between male and female flowers of the dioecious species *Rubus chamaemorus*. Bumblebees and large syrphids strongly preferred male flowers, while smaller syrphids showed less discrimination.

Comparative biology can help to address complex questions about selection for resemblance among unisexual flowers. Since wind-pollinated species have no requirements for male and female flowers to share signals, we would expect them to exhibit greater sexual dimorphism than species pollinated by animals. Data for *Leucadendron*, a dioecious African genus that has undergone multiple shifts between insect and wind pollination, support the hypothesis that selection by pollinators maintains signal similarity between males and females (Welsford et al. 2016). Strong sexual dimorphism in *Leucadendron* has been attributed mainly to sexual selection (Bond and Maze 1999). However, wind-pollinated species are more dimorphic than insect-pollinated species, particularly in the case of color and scent traits (Fig. 7.2). Emission of scent by wind-pollinated flowers may be related to defense functions and is not constrained by selection for similarity between sexes. However, resemblance of females to males is important in insect-pollinated species because the beetle pollinators obtain their main reward of pollen from male flowers only. Female flowers are therefore visited through deceit, although there have also been suggestions that female inflorescences may offer some benefits to the beetles in the form of overnight shelter (Hemborg and Bond 2005).

The trend for strong similarity in attractive signals between the sexes of dioecious species in which only one sex offers a reward has also been noted in animal-pollinated gymnosperms. In most cycads male cones act as a brood site, while female cones are visited by mistake. In general, volatile emissions of the female cones closely match those of male cones (Suinyuy et al. 2013).

In plants with traps containing unisexual flowers, insects are usually trapped during the female phase and released during the male phase. In such traps there is often strong scent production during the female stage and much weaker or absent scent production during the male phase, as the insects are encouraged to leave the trap. This violation of the rule of signal similarity between male and female unisexual flowers is an exception that proves the rule—because insects do not need to be attracted during the male phase, there is simply no selection for signals that match those of the female phase.

Resemblance among rewarding flowers

Similarities in flowers of unrelated plant species that share pollinators are apparent even to the most casual observer and were first formalized as floral syndromes by Delpino (1868–1875). Delpino's classification was developed further by Vogel (1954) and Faegri and van der Pijl (1979). Floral syndromes are simply patterns of convergent evolution and can be attributed to adaptations of plants to the sensory systems, morphology, and behavior of particular pollinator functional groups. We do not consider floral syndromes to be an example of mimicry because the resemblance among species arises incidentally through a process of adaptation to pollinators and not through selection for resemblance

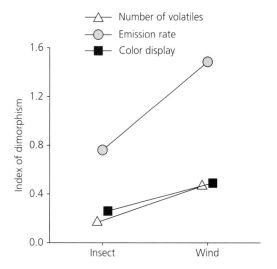

Figure 7.2 Sexual dimorphism in volatile emissions and color display is greater in wind- than in insect-pollinated *Leucadendron* species. Adapted from Welsford et al. (2016).

among organisms. However, observers have suggested that some cases of floral similarity among rewarding plant species in a community may represent a form of adaptive resemblance (Grant 1966; Brown and Kodric-Brown 1979; Powell and Jones 1983). This is a particularly contentious area of mimicry research; our purpose here is to evaluate the rather meager evidence for the idea of mimicry among rewarding plant species.

Most authors discussing examples of potential adaptive resemblance among rewarding flowers have drawn parallels with Müllerian protective mimicry (Grant 1966; Brown and Kodric-Brown 1981; Schemske 1981; Roy and Widmer 1999). The idea is that species involved in Müllerian mimicry systems will mutually benefit by reinforcing learning by operators on each other's signals. This may involve entire guilds of species ("mimicry rings") and not just pairs of species (Sherratt 2008). Müllerian mimicry involves a pattern of coevolutionary convergence, while in floral syndromes the pattern of convergence does not involve coevolution among plant species. However, the original idea of coevolutionary convergence of mimetic signals in protective Müllerian systems has been challenged, and there is now evidence that species join these mimicry systems sequentially, so that some species are the original models and others are the later mimics (Mallet 1999; Hines et al. 2011). In this revised scenario, the pattern of evolution is one of advergence rather than convergence.

Karen Grant (1966) suggested that the prevalence of the color red in particular guilds of hummingbird-pollinated plants may have arisen through a process akin to Müllerian mimicry in animals. Through her own experiments, Grant had realized that hummingbirds do not have an innate preference for red. Indeed, the prevalence of red among flowers pollinated by birds remains a long-standing problem in biology (Rodriguez-Girones and Santamaria 2004). Grant reasoned that the red color may have evolved initially in some species because of its conspicuousness to birds (and perhaps also because it was less attractive to insect nectar and pollen thieves) and that other plant species evolved red flower color simply because migratory birds were conditioned on the red color of the existing species that they encountered elsewhere on their migration

route. These plant species would in turn reinforce the conditioning of birds to prefer red flowers. In parts of South America where hummingbirds are resident and therefore have more time to sample a variety of plant species, bird-pollinated flowers have a wider range of colors. Grant's idea was adopted by Brown and Kodric-Brown (1979) who described the color resemblance in a community of hummingbird-pollinated flowers as "coevolutionary convergence." They reasoned that partitioning of pollen loads on birds would reduce competition among plant species that share hummingbird pollinators and allow them to facilitate each other's pollination by increasing the amount of nectar available in a community, and therefore the attractiveness of the community to birds. The idea of ecological facilitation among rewarding species that share pollinators was later boosted by experimental data for insect pollination systems (Thomson 1981; Laverty 1992).

The basic reasoning behind Grant's original thesis can also be applied to insect pollination systems. We know from many field and laboratory experiments that the color preferences of insects are heavily shaped by experience (Gumbert 2000; Gigord et al. 2002). It is therefore possible that similarities in color among co-flowering rewarding species can arise through selection based on conditioned preferences.

Rare rewarding plant species may benefit from deploying the signals of more common rewarding species. Such cases would represent advergent evolution and would be more similar to Batesian than to Müllerian mimicry. A good example is the similarity in flower color of the South American herb *Turnera sidoides* subsp. *pinnatifida* to co-flowering species in the Malvaceae (Benitez-Vieyra et al. 2007). What makes this potential case of mimicry particularly convincing is that the *Turnera* subspecies occurs in two color forms (which appear light orange and yellow to humans), each matching the flowers of Malvaceae in their local communities. Analysis using a bee vision model indicated that bees would be unable to distinguish the color of flowers of each form of the *Turnera* subspecies from the local Malvaceae (Box 3.1). This was supported by behavioral observations showing that bees specialized to Malvaceae mostly do not discriminate between the *Turnera* and Malvaceae flowers during their foraging bouts. One of the color forms has better pollination success when growing with the Malvaceae than is the case when it grows alone. The colors of the *Turnera* subspecies are also unusual within their lineage, suggesting a role for adaptation.

The South African orchid *Brownleea galpinii* represents another likely case of mimicry by a rewarding plant species (Johnson et al. 2003). This orchid has features that are highly derived within its lineage, including cream color and inflorescence compression, and these impart a striking resemblance to flowers of the widespread species *Scabiosa columbaria* (Fig. 7.3). The long-proboscid flies that pollinate this orchid do not discriminate between its flowers and those of the *Scabiosa* in choice experiments. The *Scabiosa* has a conserved inflorescence and flower morphology and is pollinated by many other insects across its range. It is therefore most probable that the resemblance between the orchid and the *Scabiosa* is a case of advergent evolution driven by mimicry.

In most cases of floral resemblance it remains very difficult to distinguish between floral syndromes (incidental resemblance arising from adaptation to a common pollinator) and floral mimicry (adaptive resemblance). The theory of floral syndromes predicts general convergence among flowers that share a pollinator functional group, but does not preclude the evolution of distinctive floral signals that facilitate pollinator constancy. Bees, for example, will often show constancy to a single plant species even

when many other plant species belonging to the same floral syndrome share the same community. Mimicry, by contrast, involves selection for signals that are identical among species and this, by definition (the concept of cognitive misclassification or receiver error), leads to a breakdown in constancy.

We therefore disagree with some authors who have gone as far as conflating floral syndromes and mimicry into a single phenomenon. Powell and Jones (1983, p. 312), for example, defined mimicry as a "type of convergence in which the selecting factor, or operator, is a sensory acute animal." Instead, we maintain that selection for resemblance among species in a community is a process that is distinct from the evolution of floral syndromes, and that suitable experiments should be designed to isolate this form of selection.

One way to test whether pollinator experience in a community context shapes the evolution of floral signals is to determine whether floral signals of species that share a pollinator functional group covary among communities. However, there are several caveats to this approach. Firstly, co-variation in signals among plants may reflect differences in the assemblage of pollinators among these communities. Secondly, innate color preferences of particular pollinator species may vary among populations in different communities, although the available evidence for this is rather equivocal (Raine and Chittka 2007). Another approach would be test the prediction that rewarding plants which mimic each other's signals (Müllerian mimicry) should be ecologically interdependent. This could be done by removing species from a given community and testing for the effects of this on the fecundity of other species. However, this approach has the serious caveat that mutualisms can also occur between plant species that are not mimics (Thomson 1981).

To summarize, almost all cases of floral similarity among unrelated rewarding species can be explained by convergent adaptations to the same group of pollinators—in other words, a conventional floral syndrome. It is extremely difficult to build a convincing case for floral mimicry as an explanation for similarities among rewarding species. The solution is to design better tests for other core tenets of the concept of Müllerian mimicry, including geographic co-variation in signals (mimicry

Figure 7.3 Two South African orchid species mimic the rewarding inflorescences of *Scabiosa columbaria* (center). *Brownleea galpinii* (left) is a rare rewarding species, while *Disa cephalotes* subsp. *cephalotes* (right) is non-rewarding. In both cases the orchids have highly modified phenotypes and show advergent evolution to the conserved phenotype of their model. Adapted from Johnson et al. (2003).

rings), ecological interdependence, adaptive resemblance, and cognitive misclassification by operators.

Next we consider another example of signal sharing between organisms that produce rewards, but in this case the signals are produced by organisms belonging to different kingdoms.

Fungal pseudoflowers

Endophytic fungi, floral signals, and insects

Some endophytic fungi depend on insects to transfer either gametes or infectious spores in order to complete their life cycle. As a consequence, such fungi sometimes induce the production of insect-attracting signals in their host plant. Because such signals sometimes resemble floral signals, this phenomenon is often referred to as "floral mimicry" in fungi (Batra and Batra 1985; Roy 1993; Ngugi and Scherm 2006). This ability, although investigated in only a very few species, is apparently phylogenetically widespread, being found in the rust fungi genera *Puccinia* and *Uromyces* (Basidiomycota) as well as in the Ascomycota of the genus *Monilinia*. In *Puccinia* and *Uromyces*, insects are required for sexual reproduction (i.e., transfer of spermatia acting as gametes between different fungal mating types), whereas in *Monilinia vaccinii-corymbosi*, infected *Vaccinium* (blueberry) leaves produce conidia (asexual spores) that need to be transferred to stigmas by insects. Infected ovaries subsequently produce "mummy berries," where infectious ascospores (sexual spores) are produced (Batra and Batra 1985).

Other fungi also require insects to vector spores, but do not induce "de novo" traits that resemble flowers. For example, the basidiomycete *Microbotryum violaceum* produces infectious spores in flowers of infected plants (Caryophyllaceae), and uses flower-visiting insects to transfer them from flower to flower (Carlsson and Elmqvist 1992). Although floral signals are changed upon infection by this fungus (Dötterl et al. 2009), moth pollinators have been shown to discriminate against infected *Silene latifolia* flowers (Shykoff and Bucheli 1995), thus *M. violaceum* does not classify as inducing floral mimicry because the induced changes in floral signals are seemingly not adaptive for the fungus. Another example is ascomycetes of the genus *Epichloë* that infect grasses and produce fruiting bodies called "stromata" on the host's inflorescences ("choke disease")(Bultman and Leuchtmann 2008). Stromata emit volatiles that attract specialized anthomyiid flies that transfer spermatia (gametes) (Schiestl et al. 2006); flies oviposit on stromata where their larvae feed and develop. This phenomenon thus resembles nursery pollination in plants, but both stromata and their chemical attractants are unique and do not resemble those of flowers (Steinebrunner et al. 2008). Generally, based on the different life cycles of endophytic fungi, different selection for mimicry can be expected; when insects transfer gametes, the visitation should be primarily between different mating types of fungi, i.e., insects should visit infected tissue and rather not switch to flowers where gametes may get lost (Roy 1996). Thus, perfect resemblance to flowers may not necessarily be advantageous; nevertheless, these fungi are dependent on insect visitors being drawn from the community of pollinators that visit nearby flowers. When insects transfer infectious spores via flowers, it is important for fungi that insects visit uninfected flowers after picking up the spores from infected tissue (e.g., *Monilinia*). Here, direct selection for resemblance to flowers in infected tissues can be expected.

Perhaps the most suggestive example of floral mimicry in fungi is the induction of "pseudoflowers" in host plants, a phenomenon known from some rust fungi of the genera *Puccinia* and *Uromyces* (Roy 1993, 2001). Fungal pseudoflowers are formed through a change in the developmental program of a host plant, mediated by genome-wide reprogramming of gene expression in the infected tissues (Cano et al. 2013). Pseudoflowers are characterized by rosette-like leaves at the terminal shoot that are often yellow, produce a sugar solution, and emit a sweet, flowery smell (Roy 1993; Roy and Raguso 1997; Pfunder and Roy 2000). Within the sugar solution, gametes (spermatia) are presented which are eventually picked up by insect visitors and transferred to another fungal mating type. Because pseudoflowers resemble real flowers to the human eye, produce a sugar reward, and are visited by the same insects as real flowers, their production is considered a case of Müllerian mimicry. Müllerian mimicry predicts a sharing of visitors between mimic (pseudoflower) and model (co-flowering plants), signal similarity, and an adaptive value mediated by this similarity. As discussed in the next subsection, some of these predictions are fulfilled in pseudoflowers whereas other are not, making pseudoflowers a less convincing case of floral mimicry than intuitively expected.

Are pseudoflowers Müllerian mimics of real flowers?

The pseudoflowers induced by rust fungi usually do not resemble the uninfected flowers of their host species (Roy 1993; Naef et al. 2002). This makes sense because spermatia need to reach a fungus individual of a different mating type (sex) and not be deposited on uninfected flowers (Roy 1996). So the primary selection should be on increasing constancy of insects among pseudoflowers, but also on co-opting insect visitors from flowers. For the latter task, pseudoflowers may mimic co-flowering plant species. Insect attraction to pseudoflowers induced by *Puccinia monoica* has been shown to be dependent on both color and scent (Roy and Raguso 1997). Pseudoflowers induced by this fungus are yellow, resembling the color of many co-flowering species, but lack the UV reflection that is found in some co-flowering yellow plants (Roy 1993). Even less similarity between mimic and potential models is found in pseudoflower volatiles, which differ strongly from the floral volatiles produced by typical co-flowering species as well as from the flowers of the host plants (Raguso and Roy 1998; Naef et al. 2002). Many aromatic volatiles produced by *Puccinia monoica*, for example, are not found in co-flowering species. Consistent with this finding, monoterpene biosynthesis genes were found to be downregulated in pseudoflower tissue, whereas one gene in the pathway leading the aromatic volatiles phenylacetaldehyde and 2-phenylethanol was upregulated (Cano et al. 2013).

Despite these differences in signals, pollinator insects in all systems do switch between infected tissue and uninfected flowers of host plants (Batra and Batra 1985) and/or co-flowering plant species (Roy 1993; Pfunder and Roy 2000). Thus, a perfect match in signals of mimic and model is obviously not needed to attract visitors, which in these systems are often highly generalist.

Adaptive value of similarity

Signals in Müllerian mimics should be adaptive in the sense that the mimics profit from their resemblance to their models. The likely outcome is positive density dependence of reproductive success, meaning that the more individuals (mimics and models) occur together, the more successful they should be in attracting their "pollinators" because their detectability is increased and more rewards are available. This assumption was tested for *Puccinia* and *Uromyces* rust fungi in a series of experiments conducted in the United States and Switzerland by the ecologist Barbara Roy and colleagues. First of all, Roy showed that fungi are dependent on insect visitation for sexual reproduction (Fig. 7.4) and visitors do indeed move between pseudoflowers and buttercups (*Ranunculus inamoenus*), a commonly co-flowering plant species. Pseudoflowers were even more attractive than real flowers as more switches from buttercups to pseudoflowers than the reverse were observed, and visitors stayed for longer period of time on the pseudoflowers than on the buttercups. This is likely caused by the copious amounts of nectar produced by pseudoflowers that take their resources from the host plant (Fig. 7.5). As evidence for a positive effect of the presence of the model on the mimic, the occurrence of buttercups increased visitation to pseudoflowers and vice versa. In addition, the visitation to pseudoflowers increased with their density. These findings indeed support the hypothesis of Müllerian mimicry.

In a follow-up study, however, Roy found that pseudoflowers induced by *Puccinia monoica* facilitate fly visitation to co-flowering *Anemone patens*, but that the fungus does not profit from the presence of *A. patens* (Roy 1996). In addition, fungal spermatia were lost when insects switched from pseudoflowers to *Anemone* flowers, and transfer of spermatia to *Anemone* stigmas reduced their seed set. All in all, this study suggested that the association between fungi and co-flowering *Anemone* may be competitive rather than synergistic. In general, interactions may contain both elements, thus eluding a clear-cut classification. Roy and

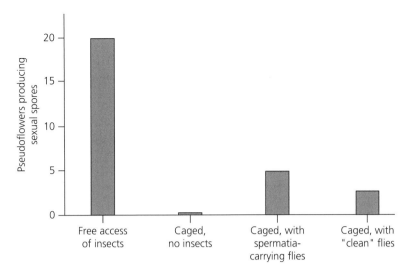

Figure 7.4 Pseudoflowers induced by *Puccinia monoica* are dependent on insect visitation in order to produce sexual spores (aeciospores). Visitation of caged pseudoflowers by flies carrying spermatia led to a lower percentage of pseudoflowers producing sexual spores, suggesting that the flies carried incompatible dead spermatia. The fact that visitation by "clean" flies led to the production of aeciospores suggests that pseudoflowers may contain more than one mating type, or that the fungus is self-compatible to some degree. Adapted from Roy (1993).

her colleagues also conducted a series of studies on the European rust fungus *Uromyces pisi*, which induces pseudoflowers in its host plant *Euphorbia cyparissias* and is dependent on insect visitation for sexual reproduction (Pfunder and Roy 2000). As was the case in *Puccinia*, visitors to the pseudoflowers moved between pseudoflowers and uninfected *Euphorbia* flowers (Pfunder and Roy 2000). In mixed arrays, however, pseudoflowers were less attractive than real flowers and thus seemed to compete with them for insect visitation. Again, these findings suggest that pseudoflowers and uninfected *E. cyparissias* flowers compete for pollinators. Competition between fungi and "real" flowers should select for pseudoflowers producing more rewards and divergent, more attractive signals, rather than a close resemblance in signaling between plant and fungus.

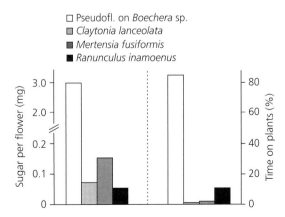

Figure 7.5 Nectar sugar production and visitation time of pollinators on *Puccinia monoica*-induced pseudoflowers and plants co-flowering in a natural habitat. Pseudoflowers produce more sugar and solicit longer visitation than do flowers in the community. Adapted from Roy (1993).

Fungal pollen mimicry

Pollen mimicry by a fungus has been suggested in the case of *Fusarium semitectum* which produces spores in the anthers of infected *Rudbeckia* (Asteraceae) flowers (Diamond et al. 2006). These spores resemble pollen and are picked up by bees visiting infected inflorescences. Although this may indeed be a case of pollen mimicry by the fungus, more data are need to support this hypothesis. One question is whether the bees actually mistake the spores for pollen and actively collect them. The available data show that bees visit infected flowers less often and spend less time on them (Diamond et al. 2006), which suggests that they are able to discriminate between spores and pollen. More behavioral assays are thus needed to support the mimicry hypothesis. In addition, mapping spore color on a phylogeny of *Fusarium* and its allies would indicate whether pollen-resembling spores have evolved in connection with the exploitation of pollen-seeking insects as vectors.

In conclusion, there is some evidence for flower-like traits induced by fungi that conform to the definition of Müllerian mimicry. For example, insects switch between infested tissue and real flowers, fungi-induced signals resemble floral signals to some degree, and in some circumstances the presence of flowers enhances visitation to fungi by insects, and vice versa. On the other hand, fungal pseudoflowers and plants sometimes compete for visitors, pseudoflowers may be more attractive than flowers, and some signals (volatiles) are quite different in the fungus "mimic" and its putative flower model. The loose signal match between mimic and model suggest a form of imperfect mimicry between fungus and flowers which may be the consequence of selection for increased attractiveness in fungi, achieved through supernormal stimulation or sensory bias (Vereecken and Schiestl 2008). Unpredictable co-flowering plant species and highly generalist spore vectors may also lead to rather diffuse selection on floral signals, because different plants with distinct floral signals can serve as models in different populations and during different seasons.

Mimicry of insect brood provisions

Many social wasps collect insect larvae in order to provision nests for their offspring. Vespid wasps that collect caterpillars locate these larvae from a distance using the volatiles that are emitted by plants upon damage by the caterpillars. A team led by the chemical ecologist Manfred Ayasse showed that signals indicative of the presence of herbivores are mimicked by some plants in order to attract predatory wasps as pollinators (Brodmann et al. 2008). The team worked on the flowers of *Epipactis helleborine* and the related species *Epipactis purpurata* that are notable for their specialization for pollination by wasps. These orchids are very seldom visited by other insects, despite producing copious amounts of nectar. The zigzag flight of the wasps toward the flowers and the results of field bioassays have confirmed that the primary signal used to attract the wasps is scent. The flowers emit the green leaf volatiles hexyl acetate, (Z)-3-hexenyl acetate, and (Z)-3-hexen-1-ol in large amounts. A synthetic mixture of these compounds was found to be attractive to *Vespula* wasps (Brodmann et al. 2008). This bizarre mimicry system is treated here as a special case, because it does not fit neatly into any of the classical systems of floral mimicry. In fact the closest resemblance is with oviposition-site mimicry involving syrphid flies, with the difference that social wasps do not oviposit close to the prey that is fed to the larvae, but rather carry them into their nest. One of the special features of this mimicry system is that the model is not a single species or group of similar species, but rather a signal arising from an interaction between two very different organisms—herbivores and their host plants.

Pollination by vespid wasps also occurs in the genus *Scrophularia* (Scrophulariaceae); investigations by the team that worked on *Epipactis* have shown that green leaf volatiles are implicated in the attraction of wasps to *Scrophularia umbrosa*, too. As in the case of *Epipactis*, wasps feed on nectar in the flowers (Brodmann et al. 2012).

It is not yet clear whether wasps cognitively misclassify the flowers of these plants as being a site where caterpillars can be collected or whether the system simply exploits the sensory bias of the wasps. Brodman et al. (2008) suggested that some of the initial interest shown by wasps toward the flowers may be related to prey finding, but that once the wasps discover the nectar, subsequent visits are made for the purpose of foraging for nectar. An interesting question that follows is whether wasps that are foraging for nectar on the flowers of *E. helleborine* are eventually able to distinguish signals of the flowers from those of herbivore-infested plants.

The Chinese orchid *Dendrobium sinense* has rewardless flowers that are also pollinated by vespid wasps. Investigations have shown that the hornet *Vespa bicolor* is attracted to the compound (*Z*)-11-eicosen-1-ol that is emitted by the flowers. This compound is an alarm pheromone of honey bees that are captured by these hornets and fed to their brood. This led Brodman et al. (2009) to suggest that the orchid mimics honey bee volatiles in order to attract hornets for pollination. Unlike the case in *Epipactis* or *Scrophularia*, the pouncing behavior of the hornets on the *D. sinense* flowers, and even eventual attempts to sting into the flower, is consistent with attempts at prey capture. Therefore, in this system cognitive misclassification of the flower by the hornet seems likely.

Male-biased pollinator assemblages

A number of recent studies have reported a strong bias toward male visitors in plants that appear to have generalized food deception. The simplest explanation is that these plants are pollinated mainly by male insects patrolling territories that include flowers likely to be visited by male insects (Johnson and Steiner 1994). Other cases appear to be more complex and are not yet well understood. A number of studies of the Mediterranean orchid *Anacamptis papilionacea* have consistently shown a strong male bias (>75%) among its bee visitors (Fig. 7.6). Vogel (1972) reported that males landing on flowers did not always extend their tongue in expectation of a food reward, suggesting that food deception was not the primary mode of attraction. Schiestl and Cozzolino (2008) found that the floral scent of this orchid is dominated by alkanes and alkenes similar to those emitted by sexually deceptive orchids, yet *A. papilionacea* does not show the pollination system specificity typical of sexual deception, and shares some pollinators and forms hybrids with *Anacamptis morio*, a species that deploys classical generalized food deception. An equally curious case is offered by the South African orchid *Disa spathulata*, which was found to be pollinated in natural populations only by males of two species of *Tetraloniella* bees, and yet plants translocated to other sites also attracted some female bees (Johnson et al. 2005). This high level of specificity is typical of a sexually deceptive species, yet the attraction of female bees in some places is not consistent with this hypothesis. Further studies of these male-biased pollination systems are required to determine whether other aspects of the life history of the bees, such as lekking behavior, are exploited by these plants, or whether they represent an initial step along the pathway to sexual deception.

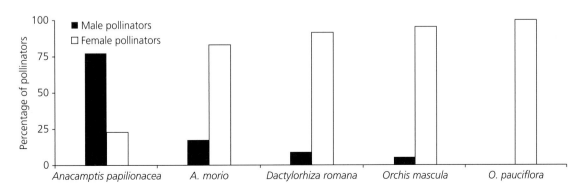

Figure 7.6 Comparison of male and female pollinators of *Anacamptis papillionacea* and other food-deceptive orchids. *Anacamptis papillionacea* has an unusually male-biased assemblage of pollinators. From Scopece et al. (2009).

Overview and perspectives

We have not included all proposed examples of floral mimicry in this book. In most cases, this is because the evidence presented in these studies is too preliminary. A case in point is the intriguing suggestion that flowers of the Japanese orchid *Cymbidium floribundum* mimic insect aggregation pheromones (Sugahara et al. 2013). This orchid is attractive to all castes of the Japanese honey bee *Apis cerana japonica* and has even been used in traditional Japanese beekeeping to attract swarms of honey bees. The flowers emit volatiles identical to those secreted by the honey bee mandibular gland, including 3-hydroxyoctanoic acid and 10-hydroxy-(*E*)-2-decenoic acid (Sugahara et al. 2013). However, the major gap in the story is that there is no convincing demonstration that honey bees actually pollinate the orchid and were therefore the agents of selection for mimicry for the chemical signature.

Miriam Rothschild's (1981) argument that many mimicry systems cannot be neatly organized into categories is highlighted by the special cases we have considered in this chapter. The value of these special cases is that they illustrate the potentially ubiquitous nature of mimicry phenomena and also the fact that many mimicry systems involve more species than just a single mimic, model, and operator.

Interspecific resemblance among unisexual flowers does not automatically qualify as mimicry, but in cases where the floral parts of one sex are modified to resemble non-homologous structures of the other sex, such as stigmas modified to resemble anthers, we feel that the resemblance is likely to be adaptive. This does challenge the conventional definitions of mimicry, even the one we have used in this book, because different sexes of a species are not usually considered to be separate "organisms," and this is certainly not the case for male and female flowers of a monoecious species.

The few well-demonstrated cases of food-source mimicry in rewarding plant species have properties that are closer to Batesian than to Müllerian mimicry. This is because one species is usually much rarer and less rewarding than the other species, and therefore more likely to be a mimic that showed advergent evolution in relation to the other species. This is part of the general problem of a mimicry spectrum. An animal which is weakly defended may show quasi-Batesian mimicry, much like a plant that is rewarding yet rare, or which offers inferior rewards. Since species pollinated by the same animals often vary markedly in their abundance and floral rewards, we think it is plausible that selection for resemblance has played some role in shaping their floral signals. However, unambiguous evidence for this scenario is currently lacking.

Because fungal pseudoflowers produce rewards and share signals with some flowers, their relationship to plants that share the same insect visitors has been described as Müllerian mimicry. The evolution of signals in these pseudoflowers may best be explained in terms of the sensory biases of insects, leading to mimicry-like signal evolution in pseudoflowers. However, it seems very unlikely that the signals of pseudoflowers and plants have coevolved in the classical Müllerian fashion.

The focus in mimicry research has tended to be on the palatability (or in the case of flowers, rewards) of the mimics, but factors such as relative abundance, geographic spread, evolutionary lability (not just of signals, but also of defenses and rewards), the relative ages of species, and learning and discrimination ability by operators are all equally important pieces of the puzzle of resemblance among species. If mimicry does occur among rewarding species, and we think it is likely that it does, then the evolution of signals needs to be studied in a community context. There is an increasing consensus that signal evolution is not only an outcome of two-way interactions between species and their predators or pollinators, but also reflects more complex interactions among multiple species in the same community.

CHAPTER 8

Future directions in floral mimicry research

Introduction

More than 200 years after the discovery of floral mimicry by Sprengel (1793), many questions about this fascinating natural phenomenon remain unanswered. Floral mimicry has traditionally been a relatively small field of research compared with animal mimicry; however, this gap is closing, with floral mimicry studies now accounting for almost a third of all studies dealing with mimicry among organisms (Fig. 1.1). For example, molecular approaches to the evolution of mimicry are well developed for animal systems, particularly butterflies, and are being used increasingly as a tool for studying plant systems. Here we review some promising future research directions in floral mimicry. We discuss insights from animal mimicry research and how these could influence future research in floral mimicry. We also identify where mimicry in plants may be the most suitable systems for studying unresolved questions relating to the evolution of mimicry in general. Each previous chapter of this book has included an "Overview and perspectives" section highlighting promising future research directions relating to the content of that chapter. Here we focus on outstanding questions that are of significance for all mimicry systems.

Mimicry and speciation

Mimicry has been suggested as a driver of plant speciation, but many details of this process remain obscure. For example, Cozzolino and Widmer (2005) proposed that floral rewardlessness may be an important factor in promoting orchid diversity, and assembled evidence from phylogenetic, ecological, and population genetic studies to support

this claim. However, it is important to discriminate between mimicry and generalized food deception (GFD) as alternative strategies for attracting pollinators to rewardless flowers (see Chapter 3). These two strategies have different implications for the role that selection plays in diversification and also for overall patterns of diversification. Strong floral isolation likely to promote speciation, for example, is only inherent in true floral mimicry, whereas GFD is often characterized by the sharing of pollinators between co-occurring plant species (Cozzolino et al. 2005). Floral ethological isolation, the lack of pollinator sharing in sympatric species, is the consequence of the high levels of specificity found in the pollination of floral mimics (Xu et al. 2011; Whitehead and Peakall 2014).

Another component of reproductive isolation in mimics is the presence of extrinsic (ecological) post-zygotic barriers. These have been highlighted in two species of *Heliconius* butterflies with different wing color patterns, where hybrids are non-mimetic and thus less protected than their parent species (Jiggins et al. 2001). Interestingly, these recently diverged species also show pre-zygotic isolation (assortative mating) and thus hybridization is rare. In floral mimics, extrinsic post-zygotic isolation has only rarely been investigated, probably because of the difficulty in reliably identifying hybrids in the field, but was found to be weak in a pair of hybridizing *Ophrys* species (Stökl et al. 2008).

Mimicry and phenotypic divergence

Another interesting but largely unexplored field is the role of the community of flowering plants on floral diversification in mimics. An example of this has been documented in South Africa, where the food

Floral Mimicry. Steven D. Johnson & Florian P. Schiestl.
© Steven D. Johnson & Florian P. Schiestl 2016. Published 2016 by Oxford University Press.
DOI 10.1093/acprof:oso/9780198732693.001.0001

mimetic orchid *Disa ferruginea* co-occurs with either orange or red flowers in geographically separated areas (Newman et al. 2012). This is likely the consequence of pollinator-mediated selection, as the pollinator prefers either the orange or the red form of *D. ferruginea* flowers in different areas, depending on the abundance of its main nectar plants. Thus in this mimicry system the flowering plant community has driven diversification by influencing the preferences of the operator. This could be a widespread phenomenon that is also relevant in other mimicry systems. In sexual mimicry, for example, the mating preferences and specificity of male insects is likely to depend on the diversity of closely related insect species in any habitat. In theory, males should be more specific (or choosy) when many other species are present, to avoid approaching or mating with heterospecific females. The consequence of this (untested) hypothesis would be that pollinators of sexual mimics should be more specific in habitats with higher insect diversity, thus selecting for stronger divergence among mimics and improving floral isolation by narrowing the signaling channels that mimics use to attract specific operators.

Another virtually unexplored question is the importance of mycorrhizae in mediating habitat preferences and phenotypic diversification in floral mimics. Mycorrhizal associations are an obligate component of establishment for almost all orchids, but the degree of dependence of adult orchid plants on a carbon supply from fungi varies (Waterman and Bidartondo 2008). Some orchids growing in forests under low-light conditions obtain large amounts of their carbon from ectomycorrhizal fungi (Gebauer and Meyer 2003), and some species have completely lost the ability to photosynthesize, having become fully dependent on a carbon supply from fungi (myco-heterotrophy). Myco-heterotrophy has evolved at least 20 times within the orchids (Molvray et al. 2000). Mycorrhizal associations were likely a key factor in the ability of orchids to colonize new habitats, such as dark shady places in forests where photosynthesis is inefficient. The ability to grow in low-light habitats may have been the pre-requisite to recruit pollinators abundant in such habitats, such as fungus gnats, a very commonly utilized group of pollinators in mimetic orchids (Phillips et al. 2014b). In addition to facilitating habitat use, fungi may also provide metabolites to their orchid partners, thus enabling the synthesis of secondary metabolites that the orchids would otherwise not be able to produce. Nothing is as yet known about this intriguing possibility.

Improving our understanding of floral signals

Floral mimicry research has traditionally focused on the significance of traits responsible for the attraction of pollinators through deceit. This is an important aspect of floral mimicry, because not all floral traits of mimics are involved in pollinator attraction and thus under selection to resemble a model. For example, a classic paper on food mimicry in *Cephalanthera rubra* has shown that despite the fact that the visual resemblance between *Cephalanthera* and its model, bellflowers (*Campanula*), is not perfect to the human eye, it is virtually identical in the visible spectrum of the pollinators, namely solitary bees of the genus *Chelostoma* (Nilsson 1983b). Similar results were found for the olfactory traits of many sexual and oviposition mimics. Nevertheless, after several decades of focus on floral traits, we still do not know all the key traits involved in floral mimicry and how pollinators respond to them. Perhaps the most striking example is the recent discovery of the key role of volatiles in the *Cephalanthera–Campanula* food mimicry system. Although the pollinator bees in this system use the UV–blue/blue color of the flowers, they also rely on specific floral volatiles—spiroacetals—that are emitted only in trace amounts (Milet-Pinheiro et al. 2015). *Cephalanthera rubra*, which mimics bellflowers and uses the same bees as pollinators, was traditionally thought to mimic only the color of its model; recent investigation has shown, however, that *Cephalanthera* emits spiroacetals, too, and that they play an important part in this mimicry system (Ayasse et al. 2015). This finding calls for a reassessment of the common view that Batesian food deception relies primarily on color, and that scent is usually unimportant in these systems. Unusual chemical communication systems have recently also been analyzed as a key element of new floral mimicry systems in *Aristolochia* and *Ceropegia* species pollinated by kleptoparasitic flies attracted by the scent of freshly killed insects

(Heiduk et al. 2015; Oelschlaegel et al. 2015). There is little doubt that an increased focus on little-investigated traits such as chemical signals will lead to a better understanding of mimicry systems, and possibly even to the discovery of new examples of floral mimicry.

Genetic bases of traits important for mimicry

The genetic and genomic bases of adaptation have become a major area of research over the past decades. Studies of the genetics of mimicry patterns in butterflies have a long tradition, enabled through the ease of crossing species and their short generation times (Punnet 1915). The increased availability of sequencing and bioinformatics tools has greatly increased the feasibility of this avenue of research (Mallet 2015). These research approaches also hold great promise for floral mimicry systems. Whereas functional studies of phenotypic traits are important for testing mimicry hypotheses, the molecular bases of those traits can inform us about the evolutionary ancestry of mimicry. Floral mimicry is a textbook example of adaptive evolution, and some of the adaptations involved in floral mimicry are among the most unusual floral structures, in terms of both structure and chemistry. Orchids of the genus *Dracula*, for example, have labella that resemble the cap of a mushroom and are pollinated by fungus gnats (Policha et al. 2016). Some sexual mimics have highly derived floral morphology, consisting of a labellum modified into a hinge and female decoy in the genera *Drakaea*, *Arthrochilus*, and *Spiculaea*, which ensures proper uptake or deposition of pollinia while the male pollinator attempts to carry away the pseudo-female (Peakall 1990) (Plate 5). Some of these orchids emit highly unusual floral scent compounds to attract their pollinators, one even representing a new class of natural products at the time of its discovery (Schiestl et al. 2003; Bohman et al. 2014). How such unique structures evolved, and from where they are derived, are some of the most intriguing questions in floral mimicry research.

A better understanding of "mimicry genes" will allow us to address questions like: Which genes have been recruited for complex floral morphologies? Are molecular mechanisms of trait evolution conserved across different taxa? Which biosynthetic pathway genes have been co-opted for unusual fragrance production found nowhere else in the plant kingdom? How many mutations were necessary, and how quickly could these adaptations evolve? Why do orchids, in particular, tend to exhibit such a stunning diversity of mimicry adaptations? How do combinations of traits evolve, and how are they maintained in the face of potential hybridization? We will briefly review some progress from studies of animal mimicry as well as non-mimetic plant systems to highlight potential avenues for genetic studies in floral mimicry research, keeping in mind the constraints imposed by the life history of many plants that exhibit floral mimicry (Schlüter and Schiestl 2008).

The genetics of traits in organisms can be investigated through two primary but distinct approaches, namely classical quantitative genetics and molecular genetics (Galliot et al. 2006). Quantitative (or ecological) genetics focuses on phenotypic traits in the framework of their genetic basis, with genetic architectural features such as linkage and pleiotropy being important topics. Molecular genetics, on the other hand, focuses on genes, their molecular properties, and their roles within an organism. Combinations of classical quantitative and molecular genetics, such as the identification of individual genes contributing to a given trait locus, are still challenging, but hold great promise for the future (Klahre et al. 2011; Kunte et al. 2014; Nishikawa et al. 2015).

Although both quantitative and molecular approaches may be very useful in plant mimicry research, they ask distinct questions and give different insights. Even the full sequencing of a plant genome, for example, will not be informative about allelic diversity in natural populations and how alleles are being recombined during sexual reproduction. A quantitative approach such as quantitative trait locus (QTL) mapping, on the other hand, does not necessarily provide information about what kinds of genes contribute to any trait. Therefore, given the often substantial time and monetary investments needed, one must carefully choose the approach that is most suitable for a given organism and the questions being addressed. Unfortunately, quantitative genetics approaches are severely limited in mimetic plant species due to their long generation times and the difficulty of cultivating plants in a greenhouse. Typically, quantitative genetics requires the creation of hybrid populations, to ensure the segregation of traits, or multiple generations of artificial selection. This in itself is not a problem as many mimics, particularly those in the Orchidaceae, readily form artificial hybrids. However, floral mimicry is typically found in perennial plants, often with long generation times of typically several years. In contrast, *Heliconius* butterflies have generation times of around a month, and thus the results of crossing experiments can be seen very quickly. Long generation times often prohibit the production of flowering F_2 or even F_1 hybrids within useful timeframes. Additionally, many species, especially orchids, are difficult to grow from seeds because germination may require mycorrhizae and seedlings have low survival rates. Such constraints are most likely the reason why quantitative genetics has not yet been applied to floral mimics.

Molecular genetic approaches and genome sequencing, while being less dependent on growing large numbers of plants, require molecular tools (e.g., for the functional characterization of genes/enzymes) that often need to be specifically adapted to a given system (Schlüter and Schiestl 2008). Nevertheless such tools are now more readily available in non-model organisms (Hsiao et al. 2011). Molecular genetics (combined with ecological/evolutionary approaches) therefore seems to be the most promising approach currently available for learning more about the genetics of floral mimicry (Schlüter et al. 2011; Sedeek et al. 2014). We shall now discuss some research progress and new tools for unravelling the molecular genetics of floral mimicry. Quantitative genetics aspects are discussed in more detail later in "Mechanisms of evolution in mimicry."

Molecular genetics of floral mimicry

Perhaps the most notable progress in the identification of floral mimicry genes has been the identification of the genes responsible for the production of female-mimicking scent compounds in sexual mimics within the genus *Ophrys*, using the candidate gene approach. In these orchids, plants produce alkenes of different chain lengths and double-bond positions to mimic female-produced alkenes that act as sex pheromones. The key enzymes determining the production of alkenes with given double-bond positions are desaturases that introduce a double bond at a specific position in fatty acids, the precursor compounds of alkenes (Fig. 8.1). Interestingly, desaturases in plants and insects are not homologous and use different carrier molecules (Shanklin and Somerville 1991; Schlüter et al. 2011). Fatty acid desaturases in plants are encoded by the so-called *SAD* genes, and several of them have now been cloned and functionally characterized in various *Ophrys* species (Schlüter et al. 2011; Xu et al. 2012). The expression of these genes varies in a species-specific way and matches the expected patterns of alkenes produced by the flowers. *SAD* genes are also under pollinator-mediated divergent selection (Xu et al. 2012). The mechanism of regulation of *SAD* gene expression is currently unknown, although studies in hybrids suggest that both *cis-* and *trans-*mediated regulation may be important. Variation in regulatory genes, however, may be the key to switching models among sexual mimics. They may act in a fashion similar to the regulatory *dsx* gene in mimetic *Papilio* butterflies that controls the expression of genes that produce mimetic wing patterns (Kunte et al. 2014). Regulatory genes in sexual mimics may ultimately also determine mechanisms of speciation, as sexual mimicry usually comes with strong floral isolation between different forms mimicking different models. Alternatively, structural genes under strong selection or *cis-*regulatory variation controlling structural genes may be key factors for switching odor bouquets (Xu and Schlüter 2015). Some progress has also been achieved in identifying the genes that regulate thermogenesis in oviposition-site mimics (Onda et al. 2015). Thermogenesis depends on a mitochondrial uncoupling gene or an alternative oxidase. These genes have indeed been found to be expressed in different Araceae species with thermogenesis (Chapter 6) (Ito and Seymour 2005).

Genome sequencing

The sequencing of full genomes is progressing rapidly; more and more sequenced genomes of non-model species are now available. A fully sequenced genome provides a wealth of information on genome structure and the genes themselves, and allows

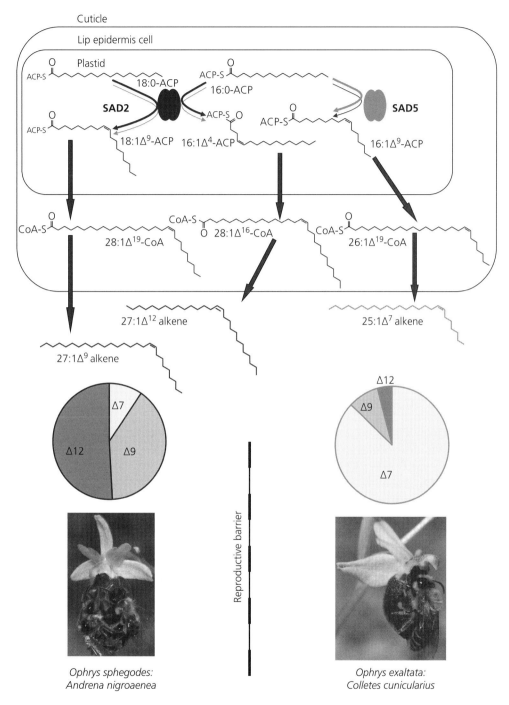

Figure 8.1 Schematic drawing of the biosynthetic pathways leading to the different alkenes involved in pollinator attraction in *Ophrys sphegodes* and *Ophrys exaltata*. The key enzymes are two different desaturases, SAD2 and SAD5, which introduce double bonds into fatty acids. The end products, alkenes with different double-bond positions, lead to the attraction of different pollinators, *Andrena nigroaenea* and *Colletes cunicularius*, respectively. Adapted from Schlüter et al. (2011).

researchers to allocate tags of sequenced DNA within the genome as well as detailed genome-wide association studies (GWAS). GWAS investigate the association of traits and single nucleotide polymorphisms (SNPs) and can be useful in the search for functional genes in natural, variable populations of organisms.

In *Heliconius* butterflies, genome sequencing has revealed that species have exchanged genes responsible for mimicry color patterns during their evolutionary history (Dasmahapatra et al. 2012). Even a few hybridization events can lead to such an adaptive exchange of genes, suggesting that hybridization can play an important role in the evolution of mimicry. In *Papilio polytes*, a butterfly polymorphic for Batesian mimicry, genome sequencing has led to a better understanding of the molecular mechanisms controlling polymorphic mimicry patterns (reviewed under "What are the origins and evolutionary patterns of mimicry genes?") (Nishikawa et al. 2015). Currently, genome-wide molecular data are not available in any plant species exhibiting floral mimicry; however, several orchid genomes have been sequenced and more sequencing is in progress.

Although the full genome is not yet available, the orchid genus *Erycina* may be the most interesting for mimicry research because of its close relationship to *Oncidium*, a genus containing several Batesian mimics of rewarding Malpighiaceae flowers (Carmona-Diaz and Garcia-Franco 2009). The chloroplast genome of *Erycina pulsilla* has recently been sequenced (Pan et al. 2012) and transcriptome data are available for this species (Chou et al. 2013). *Erycina* has a low chromosome number, a small genome, and a relatively short generation time, making this genus—along with the related *Oncidium*—a promising model for genomic approaches to floral mimicry (Pan et al. 2012). An alternative to full genome sequencing is transcriptome sequencing, which usually requires less sequencing (because the transcriptome is smaller than the genome) and gives an overview of genome-wide gene expression, potentially leading to the discovery of interesting functional genes.

Special tools for functional molecular investigations

For many aspects of molecular genetics, tools are required that often only work in a taxon-specific way

and are time-consuming and expensive to develop. Thus it is no big surprise that their availability is currently limited to model plants such as *Arabidopsis thaliana*. Some tools have successfully been adapted to commercially important orchid genera such as *Phalaenopsis, Oncidium, Dendrobium*, and *Cymbidium* (Hsiao et al. 2011). Of these, only *Oncidium* contains species with likely Batesian mimicry (Papadopulos et al. 2013). Molecular tools are needed to characterize the function of a given gene. Most elegantly, genes can be silenced and the effect on the phenotype examined. Such genetically modified plants can also be used for testing the importance of a given trait for pollinators (Kessler et al. 2015). A powerful tool that has been successfully applied in orchids is virus-induced gene silencing (VIGS). VIGS has the advantage of not requiring trans-generational cultivation of transformed plants. By this method, a fragment of the gene of interest is cloned into a plant virus, with which plants are then infected: This results in the suppression of viral RNA by the plant's own anti-viral defense mechanisms, and concomitant co-suppression of the target gene. Through this method, several homologous transcripts are typically co-silenced, which can be problematic when the experiment requires high specificity. The method has been successfully used with the *Cymbidium* mosaic virus, an orchid virus that can be inoculated via *Agrobacterium tumefaciens* (Hsieh et al. 2013). Finally, in the future, tools for genome editing such as CRISPR (clustered regularly interspaced short palindromic repeats) may become available for non-model plants, opening up extraordinary possibilities for studying the molecular basis of mimicry (Schiml and Puchta 2016).

Mechanisms of evolution in mimicry

One of the oldest questions in mimicry research relates to the evolution of conspicuous traits like bright coloration, typically found in protective mimicry, in which the warning colors of distasteful models are imitated by palatable mimics. During the evolution of such warning coloration, an organism changes from initially being cryptic to becoming highly visible. The gradual evolution of mimicry from crypsis thus likely involves an adaptive valley, as the organism loses the protection that it gained through crypsis (Harvey and Paxton 1981). This raises the question as to whether mimicry evolves gradually or in a saltational fashion. Relevant for this qestion are both the genetic architecture of mimicry traits and the sensory ecology of operators. Because multiple traits within a mimicry system are unlikely to have evolved simultaneously, a two-step process of mimicry evolution has been suggested, whereby an initial saltational mutation achieves approximate resemblance to a distasteful model (or resemblance in few key traits), followed by gradual evolution that fine-tunes the similarity (Gamberale-Stille et al. 2012). According to this model, categorization by the operator based on few traits is necessary, because the first step may only establish similarity in the few traits that the operators use for categorization. The two-step model is largely supported by insights into the genetic mechanisms of color pattern regulation in mimetic butterflies, where a few loci with a large effect were found, in addition to small-effect regulatory loci that fine-tune the expression of the mimetic pattern (Kronforst and Papa 2015).

The two-step model may also apply to floral mimicry. Categorization by pollinators (i.e., the use of a few traits to decide on the identity of a flower) has been shown, for example, in bumblebees and honey bees that are well known to respond to artificial flowers that emit only a subset of the attractive signals of real flowers (Raguso 2001). Pollinators can be attracted to color signals in the absence of scent signals, and vice versa. In addition, individual volatile compounds may be as effective or nearly as effective as the whole bouquet (Plepys et al. 2002; Dötterl et al. 2006). A recent study of *Brassica rapa* showed that bumblebees use individual compounds within the total scent bouquet that provide honest signals about the amount of reward present (Knauer and Schiestl 2015). Thus, an initial similarity in key signals that pollinators use to find rewarding flowers may deliver enough visits to ensure reproductive success, and fuel selection for improved resemblance in other traits. Evidence from mimicry systems also supports categorization by pollinators; for example, Shuttleworth and Johnson (2010) have shown that sulfur-containing volatiles (typically emitted by carrion) can initially attract carrion flies to flowers

that bear no additional similarity to carrion. Therefore, emission of these sulfides could represent the saltation for mimicry, while evolutionary fine-tuning would explain why most carrion flowers emit a combination of several different biochemical classes of carrion volatiles as well as visual and even tactile signals associated with carrion (Urru et al. 2011). In sexual mimics, a single volatile compound may be enough to attract sexually aroused male pollinators (Schiestl et al. 2003), but sexual mimics also typically mimic a combination of signals, such as the scent, shape, color, and hairiness of a female insect (Sedeek et al. 2014; de Jager and Peakall 2016).

The genetic architecture of mimicry

Because it is not individual genes but their combinations that control a complex adaptation such as mimicry, quantitative approaches to the genetic architecture of mimicry promise deeper insights into the evolution of this phenomenon. For example, pleiotropy may enable but also constrain the evolution of mimicry. Pleiotropy between leaf and flower pigmentation has been suggested to impact on the evolution of floral color (Armbruster 2002; Rausher 2008). On the other hand, evidence for pleiotropy between scent and color is weak (Majetic et al. 2007; Salzmann and Schiestl 2007). Mimicry genes may also be linked because of tight selection on combinations of traits, such as different scent compounds, color, and trichomes. Mutants with one of these traits that is not properly developed will lose their similarity to the model and thus have reduced fitness. An exception occurs in cases where a reduced similarity between model and mimic actually enhances fitness. This can be one of the pathways leading to the evolution of imperfect mimicry (Vereecken and Schiestl 2008), but even here similarity must remain within the perceptual frame of the operator to be effective.

Much progress has been achieved in the genetics of mimicry traits in *Heliconius* butterflies that engage in Müllerian mimicry (Kronforst and Papa 2015). In many of these species, color patterns are controlled by a few loci of large effect. These loci have classically been interpreted as "supergenes," consisting of several tightly linked genes (Mallet and Joron 1999). Besides these loci of large effect, variation in mimicry patterns is controlled by the action of small-effect modifier loci. Similarly, genetic dissection of floral traits comprising pollination syndromes in *Petunia* led to the discovery of a few QTLs of large effect (Galliot et al. 2006). Such loci of large effect, however, may represent several linked genes or a few genes with pleiotropic effects on multiple traits. Linkage does indeed seem to play an important role in the trait combinations that underlie pollination syndromes. For example, Hermann et al. (2013) found strong linkage in loci that are important for traits involved in pollination syndromes in *Petunia*. A comparison of these loci within the Solanaceae suggests that they have evolved to become physically clustered, perhaps by selection against recombination of the traits. Such linkage provides the basis for maintenance of trait combinations in the face of gene flow. In earlier studies, Klahre et al. (2011) detected two QTLs for scent production in *Petunia*, localized on different chromosomes, one of which was shown to be the transcription factor ODORANT1. Similarly, differences in stamen and pistil lengths among two *Petunia* species with different pollination syndromes are controlled by two QTLs (Hermann et al. 2015). There are no equivalent data for floral mimics, but these findings for *Petunia* give a sense of what may yet be discovered about the genetic architecture of floral mimicry traits.

The key limitation in many floral mimicry plants is undoubtedly their difficult germination and growing conditions and their life history involving late flowering. Progress in this field will thus depend on improvements in the cultivation of plant mimics in the greenhouse, or the identification of model systems that are easy to cultivate and have short generation times. Oviposition-site mimics in the genus *Aristolochia* may be promising candidates in this respect.

What are the origins and evolutionary patterns of mimicry genes?

Mimicry often involves traits that seem to be unique within a clade of closely related species and it is not always easy to identify the precursor traits. For example, the scent of sexual mimics in the genus *Chiloglottis* is unique and has up to now been found only in closely related orchids that also engage in sexual mimicry (Peakall et al. 2010). Similarly, the floral morphology of the sexual mimics *Drakaea, Arthrochilus*, and *Spiculaea* with their hinged labella is without parallel in the plant kingdom. Insights into how such stunning adaptations evolved can be gained from the evolutionary trajectories of mimicry genes. For example, in *Papilio polytes* butterflies some females are Batesian mimics of toxic species, whereas others are non-mimetic. Earlier quantitative studies led to the identification of the so-called H-locus which controls the mimicry wing pattern in this species. The H-locus was originally thought to be a supergene, consisting of several tightly linked genes (Mallet 2015). A recent report, however, showed that the H-locus is actually a single regulatory gene, called *doublesex* (*dsx*), that controls sex determination and aspects of sex-specific differentiation in other animals (Kunte et al. 2014). This finding is surprising, because it seems unlikely that a conserved transcription factor with important basal function can be co-opted for mimicry. The most plausible explanation seems to be that the modularity of the encoded protein, with only some protein domains being required for the conserved function, allows for multiple functions within this single gene (Nishikawa et al. 2015). Similarly, the molecular mechanisms of scent production in the sexual mimic *Ophrys* are relatively well understood and give insights into how genes coding for insect sex pheromones can evolve in plants. In this system, *SAD* genes coding for desaturases are of key importance, but *SAD* genes in *Ophrys* also have a conserved, housekeeping function in the plant's physiology, namely the production of essential unsaturated fatty acids. Here, gene duplications and subsequent independent diversification of *SAD* genes allowed for maintenance of the housekeeping function in some gene copies while others evolved to function in chemical mimicry (Schlüter et al. 2011).

The power of using adaptive loci for reconstructing trait evolution has been shown in two (Müllerian) *Heliconius* mimics that have co-opted the same gene controlling red wing pattern, the *optix* locus (Hines et al. 2011). Whereas phylogenies using (neutral) mitochondrial DNA supported multiple origins for red wing patterns within two species of *Heliconius*, a phylogeny of the *optix* locus supported that red wing pattern evolved only once in each *Heliconius* species (Fig. 8.2). Because of the obvious dissociation between loci that are unlinked (and even those with some degree of linkage) to the mimetic trait in recently diverged species, the history of adaptive change is best reconstructed using genes that code for the trait directly. Patterns of variation also suggest that *Heliconius erato* evolved the red-banded forms first, and *Heliconius melpomente* adverged toward them later, giving interesting insights into the relative timing of mimicry evolution in two species. In floral mimics, with their plethora of adaptive traits, reconstruction of their evolutionary patterns would shed light on the speed and number of independent origins of those traits. Mapping traits onto classical phylogenies is informative in terms of the relative timing of trait evolution in deep time (Schiestl and Dötterl 2012), but for recently radiated systems, functional genes

would provide better insights. The pattern and timing of trait evolution would be of interest, for example, in plant groups with the same mimicry systems. In oviposition mimics, it could be asked whether production of sulfur compounds evolved multiple times or only once. In Australian sexual mimics, the evolutionary patterns of pyrazines and chiloglottones could be tracked. Next, we consider the related question whether independent lineages co-opt the same genes for mimicry evolution.

Convergent evolution at the molecular level

Unrelated organisms that mimic the same models often exhibit patterns of convergent evolution (Figs 5.3 and 6.9). The molecular bases of such convergence have recently been identified in several organisms, leading to sometimes surprising insights. Broadly speaking, similar phenotypes can evolve in closely related species by different molecular mechanisms, but the same molecular mechanisms can also be responsible for similar phenotypes in distantly related organisms (Arendt and Reznick 2008). The latter phenomenon is of relevance for mimicry, and several instances of it are known. For example, the mechanism creating yellow, black, and red pigments is highly conserved across different *Heliconius* species, and the specific genes involved have similar functions in *Drosophila* (Ferguson et al. 2011). Shared molecular mechanisms were also found to be responsible for the evolution of floral traits such as color. In *Ipomea* the same mutation in the *cis*-regulatory region of a gene led to a shift from blue to red flower color in different species (Des Marais and Rausher 2010). Evolution of red coloration in *Ipomea* is caused by an independent but molecularly identical loss-of-function mutation in the anthocyanin pathway. Thus the same molecular mechanisms may cause multiple independent color shifts (Streisfeld and Rausher 2009). In floral mimicry, different mimics having the same model, and even mimics and their models, may co-opt the same molecular mechanisms for evolving similar traits. In mimetic coral snakes, for example, it has been shown that the same pigments are involved in convergent evolution of aposematic coloration, indicating the possible use of homologous genes (Kikuchi et al. 2014). On the other hand, models and mimics

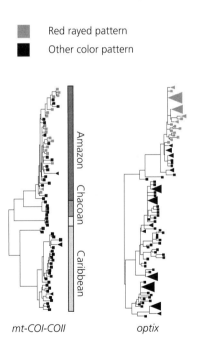

Red rayed pattern

Other color pattern

Amazon

Chacoan

Caribbean

mt-COI-COII

optix

Figure 8.2 Phylogenetic relationships of different races of *Heliconius erato* reconstructed using mitochondrial (neutral) markers (left) and the locus responsible for red-rayed wing pattern, *optix* (right). Whereas mitochondrial markers structure the population according to geography, *optix* structures mostly according to wing pattern and suggests a single origin for the red-rayed pattern in *H. erato*. Adapted from Hines et al. (2011).

may be too different to use the same mechanisms. In oviposition-site mimicry, for example, the model is not an organism but rather a by-product, such as the signals emitted as a product of bacterial activity in decaying flesh or dung. In the case of floral sexual mimicry, where models are insects, homologous molecular mechanisms for signal production seem unlikely. Indeed, desaturases responsible for alkene production, the attractive signal in *Ophrys* mimics, are not homologous in plants and insects (Shanklin and Somerville 1991).

Developmental biology in floral mimicry

Developmental processes are of fundamental importance for morphological diversification. In floral mimicry, specific morphologies often play a key role, either through signaling or mechanical function. The molecular bases of developmental processes are thus of great relevance for understanding the coding and evolutionary trajectory of certain mimicry traits. As a prime example, orchids are characterized by their floral (perianth) structure consisting of three outer tepals (sepals), two inner tepals (petals), and one modified petal, the lip (labellum). In general, the lip has been considered a key innovation in orchids, as it provides the basis for specific pollinator attraction and is itself highly diversified (Mondragón-Palomino and Theißen 2008). The morphology of the orchid labellum often provides the basis for floral mimicry, for example by forming the structure that actually mimics the model, or by providing mechanical support for ensuring pollination through the action of the pollinator. Visual and olfactory signaling by the lip is often of key importance in mimicry.

Understanding the molecular basis of lip formation in orchids has been the focus of considerable research over the past few decades. Among floral developmental genes, various classes of MADS-box genes have been suspected to play a key role in orchid lip formation (Mondragón-Palomino and Theißen 2008). MADS-box genes are transcription factors, encoding small proteins that bind to DNA, and thus regulate the transcription of other genes. These genes appear to have undergone multiple duplications at the onset of orchid diversification (Hsu and Yang 2002; Mondragon-Palomino et al. 2009).

These duplications are thought to have enabled the acquisition of new functions in these regulatory genes. Several expression studies in MADS-box genes have suggested a combinatorial model, an "orchid code" for tepal identity and lip formation, with the combinatorial expression of different MADS-box genes underlying organ specification (Mondragón-Palomino and Theißen 2008; Chang et al. 2010; Aceto and Gaudio 2011; Mondragon-Palomino and Theißen 2011; Acri-Nunes-Miranda and Mondragon-Palomino 2014). Recently, Hsu et al. (2015) proposed a refinement of the orchid code, which involves competition between two protein complexes encoded by four MADS-box genes each (AP3/AGL6 homologues). The SP (sepal/petal) complex specifies sepal/petal formation and the L (lip) complex is required for lip formation. The relative expression of these complexes leads to either more sepal/petal-like or more lip-like organs. Homeotic floral mutants of *Oncidium* that form lip-like structures instead of petals indeed show expression of the L complex in petals. Using VIGS in *Oncidium* and *Phalaenopsis*, one of the genes in the L complex was silenced, leading to a reduction in expression of the L complex. These VIGS plants produced smaller lips with sepal/petal-like structures, supporting the idea that reduced expression of the L complex leads to a shifting of the balance toward the SP complex, resulting in the formation of sepal/petal-like lip structures. The genes forming the L and SP complexes will be central to future studies of mimicry adaptations that include modifications of the lip.

Another interesting example of floral developmental mechanisms important in mimicry is the flowers of *Gorteria*, one of only two known examples of the evolution of sexual mimicry outside of the orchids (see Chapter 5 for more details on this system). Inflorescences (capitula) of *Gorteria* consist of individual flowers called florets, of the ray and disc type. Elaborated mimicry spots on the ray florets play a key role in deceiving the fly pollinators. These spots do not appear on each floret but only in a subset of them, sometimes in a seemingly random fashion. Closer inspection, however, showed that the angle between florets with spots is always 137.5 degrees. This is because the florets with spots are only initiated during a short time within floret initiation; because of spiral phyllotaxy,

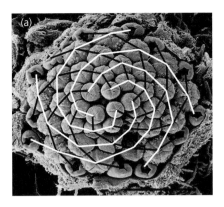 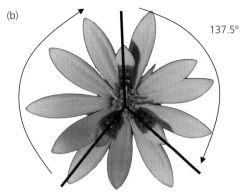

Figure 8.3 In *Gorteria diffusa*, successive florets are initiated 137.5 degrees apart from each other on the inflorescence meristem, forming a spiral phyllotactic pattern. (A) Scanning electron microphotograph of an early inflorescence bud, showing the spiral phyllotaxy. Black and white lines indicate the right- and left-turning parastichies (spiral lines of primordia), respectively. (B) Spots develop on ray florets that are developmentally sequential, resulting in a set pattern of spots, which are always 137.5 degrees apart. Adapted from Thomas et al. (2009).

consecutive florets initiate at 137.5 degrees to each other (Fig. 8.3). This simple developmental mechanism leads to a quasi-random arrangement of the spots on the inflorescence. The actual number of spots per capitulum also varies, likely according to the strength of the molecular developmental signal initiating spots. These developmental mechanisms make the mimicry of resting flies appear more natural, and thus perhaps function more efficiently (Thomas et al. 2009).

Overview and perspectives

Although this chapter has focused chiefly on molecular aspects of future research, we believe that a combination of ecological approaches (covered mainly in the preceding chapters) and molecular work will enhance our understanding of floral mimicry in the future in the most significant way. We do not want to give the impression that the ecology of mimicry is already well understood and that molecular issues are the only frontier in mimicry research. Indeed, there are many putative cases of mimicry which remain unresolved because of lack of knowledge about the identity and behavior of the

operators and the ecological dependence of mimics on models. Mimicry is essentially an interaction among organisms and will therefore always require a component of field biology. We believe that it is the community of flowering plants with its potential models, mimics, and operators that ultimately shapes selection on mimicry traits. Network approaches, recently applied in many studies of pollination biology, could also be useful for studying mimicry systems, particularly in those systems that are not very specific and show variable pollinator assemblages over their geographic ranges.

However, ecology alone cannot provide all the answers, and we advocate a multidisciplinary approach to the study of mimicry. Molecular studies in floral mimics are still in their infancy, but hold great promise. Identification of additional "mimicry genes" will provide insights into the physiological mechanisms and evolutionary patterns of mimicry traits. Quantitative approaches will show how genetic architecture enables, maintains, or constrains floral mimicry, while developmental genetics will reveal how the floral Bauplan can be modified in manifold ways, as exemplified by mimicry.

Epilogue

Is floral mimicry merely a botanical oddity, or has it instead been a major factor in the evolution of flowers? As we worked on this book we became convinced that floral mimicry and other forms of pollinator deception are not rare phenomena restricted to the orchids and a few other highly derived plant families but are, in fact, widespread among plants. This conclusion should not really have come as a surprise, given that species interactions are often exploitative.

The history of relationships between plants and flower-visiting animals has involved numerous shifts between mutualisms and various forms of deception. These shifts are much more complex than simple losses and gains of floral rewards, or changes from legitimate visitation to resource robbing by flower-visiting animals; they have also involved more subtle forms of deception, such as the deployment of pollen-imitating nectar guides in rewarding flowers. Mimicry is a key aspect of the evolutionary dynamic of shifts between mutualism and deception. Because some behavioral phenomena involved in the evolution of mimicry, such as receiver bias, are also important for the evolution of other pollination systems, studying mimicry can help us to understand how plant–pollinator interactions evolve in general.

Mimicry exemplifies the importance of the community context for trait evolution. Signal traits of flowers have traditionally been explained in terms of the sensory apparatus of pollinators, but the existence of mimicry reminds us that floral signals also need to be interpreted in a community context. Mimicry is a strategy whereby organisms exploit the behavioral responses of operators to the phenotypes of other organisms in the same community. A strange truism is that we often know far more about

the mimics than we do about the models and their interactions with the operator. More effort needs to go into understanding the interactions between operators and models, particularly in oviposition-site and sexual mimicry systems. However, the scale of this task is daunting, given the hundreds of thousands of insect species that visit flowers.

Another issue that emerged as a major limitation on our ability to understand the evolution of mimicry systems is that we don't have enough information about the role of innate versus learned preferences by operators. The reason why this matters is that theoretical models for the evolution of mimicry tend to incorporate a learning process. If the responses of the operators to phenotypes are usually hard-wired, then some of the components of these theoretical models, such as the role of frequency dependence, may need to be revised. In this respect, theoretical models that are based on signal detection theory may ultimately prove more useful than those involving a learning process. Having said this, it still seems hard to imagine that the detailed similarity in visual patterns that characterize some protective mimicry systems, such as those in butterflies and frogs, could reflect selection entirely through a hard-wired response by operators. On the other hand, hard-wired preferences of food-seeking pollinators for particular flower colors and scents seem plausible, raising questions about whether floral syndromes and floral mimicry are really so different after all.

Ecological specialization in pollination systems is a topic that has generated much discussion, but most of this has been centered on the measurement of specialization and its role in ecological flower-visitor networks. Far more attention needs to be placed on understanding the evolution of specialization

Floral Mimicry. Steven D. Johnson & Florian P. Schiestl.
© Steven D. Johnson & Florian P. Schiestl 2016. Published 2016 by Oxford University Press.
DOI 10.1093/acprof:oso/9780198732693.001.0001

itself. Given that many mimicry systems are highly specialized, often involving a single operator species, they are ideal for investigating how and why specialization evolves, and the factors that constrain it. There is some evidence that floral specialization in mimics confers advantages for male mating success by enhancing pollen transfer efficiency, but these data exist for only a limited number of orchids, and the link between mimicry and pollination success needs to be evaluated for additional plant groups. A powerful but neglected idea for conceptualizing the evolution of floral specialization is the notion of trade-offs in the deployment of signaling traits. It seems possible that the high levels of specialization in sexual mimicry result from trade-offs between traits that attract different pollinator species, though experiments are needed to test whether trade-offs generally underlie the specialization that we observe in floral mimicry systems.

Biological science involves a process of observation, then the formation of a hypothesis, then experimentation, followed by further observation, and so on. This may be seen as a positive spiral process, a kind of iterative search for truth, which is ongoing and central to the study of mimicry. In many of the cases that we have examined in this book, the science has not proceeded beyond the initial stages of observation and hypothesis formation. We need more experiments and more critical reflection on existing experiments in order to convert the dozens of speculative papers that we have read into solid case studies. If another version of this book is written 20 years from now, we have no doubt that many of the questions we have raised here will have been answered, and also that a new set of questions about mimicry will have emerged to guide the next phase of research into this intriguing topic.

References

Aak, A. and Knudsen, G. K. 2011. Sex differences in olfaction-mediated visual acuity in blowflies and its consequences for gender-specific trapping. *Entomologia Experimentalis et Applicata* **139**, 25–34.

Aceto, S. and Gaudio, L. 2011. The MADS and the beauty: genes involved in the development of orchid flowers. *Current Genomics* **12**, 342–356.

Ackerman, J. D. 1986. Mechanisms and evolution of food-deceptive pollination systems in orchids. *Lindleyana* **1**, 108–113.

Ackerman, J. D. and Carromero, W. 2005. Is reproductive success related to color polymorphism in a deception pollinated tropical terrestrial orchid? *Caribbean Journal of Science* **41**, 234–242.

Ackerman, J. D., Cuevas, A. A., and Hof, D. 2011. Are deception-pollinated species more variable than those offering a reward? *Plant Systematics and Evolution* **293**, 91–99.

Ackerman, J. D., Melendez-Ackerman, E. J., and Salguero-faria, J. 1997. Variation in pollinator abundance and selection on fragrance phenotypes in an epiphytic orchid. *American Journal of Botany* **84**, 1383–1390.

Acri-Nunes-Miranda, R. and Mondragon-Palomino, M. 2014. Expression of paralogous SEP-, FUL-, AG- and STK-like MADS-box genes in wild-type and peloric *Phalaenopsis* flowers. *Frontiers in Plant Science* **5**, 1–17.

Ågren, J. and Schemske, D. W. 1991. Pollination by deceit in a Neotropical monoecious herb, *Begonia involucrata*. *Biotropica* **23**, 235–241.

Ågren, J., Elmqvist, T., and Tunlid, A. 1986. Pollination by deceit, floral sex ratios and seed set in dioecious *Rubus chamaemorus* L. *Oecologia* **70**, 332–338.

Ågren, L., Kullenberg, B., and Sensenbaugh, T. 1984. Congruences in pilosity between three species of Ophrys (Orchidaceae) and their hymenopteran pollinators. In: *The Ecological Station of Uppsala University on Öland 1963–1983. Selected Works From 1973–1983* (ed. by Kullenberg, B., Bergström, G., Svensson, B. G., Tengö, J., and Ågren, L.). pp. 15–25. Uppsala: The Royal Society of Sciences of Uppsala.

Aigner, P. A. 2001. Optimality modeling and fitness trade-offs: when should plants become pollinator specialists? *Oikos* **95**, 177–184.

Albre, J. and Gibernau, M. 2008. Reproductive biology of *Arum italicum* (Araceae) in the south of France. *Botanical Journal of the Linnean Society* **156**, 43–49.

Alcock, J., et al. 1978. The ecology and evolution of male reproductive behaviour in the bees and wasps. *Zoological Journal of the Linnean Society* **64**, 293–326.

Alexandersson, R. and Ågren, J. 1996. Population size, pollinator visitation and fruit production in the deceptive orchid *Calypso bulbosa*. *Oecologia* **107**, 533–540.

Amich, F., Garcia-Barriuso, M., and Bernardos, S. 2007. Polyploidy and speciation in the orchid flora of the Iberian Peninsula. *Botanica Helvetica* **117**, 143–157.

Anderson, B. and Johnson, S. 2006. The effects of floral mimics and models on each others' fitness. *Proceedings of the Royal Society B: Biological Sciences* **271**, 969–974.

Anderson, B., Johnson, S. D., and Carbutt, C. 2005. Exploitation of a specialized mutualism by a deceptive orchid. *American Journal of Botany* **92**, 1342–1349.

Anderson, G. J. 1979. Dioecious *Solanum* species of hermaphroditic origin is an example of a broad convergence. *Nature* **282**, 836–838.

Angioy, A. M., Stensmyr, M. C., Urru, I., Puliafito, M., Collu, I., and Hansson, B. S. 2004. Function of the heater: the dead horse arum revisited. *Proceedings of the Royal Society B: Biological Sciences* **271**, S13–S15.

Aragón, S. and Ackerman, J. D. 2004. Does flower color variation matter in deception pollinated *Psychilis monensis* (Orchidaceae)? *Oecologia* **138**, 405–413.

Arendt, J. and Reznick, D. 2008. Convergence and parallelism reconsidered: what have we learned about the genetics of adaptation? *Trends in Ecology and Evolution* **23**, 26–32.

Arguello, J. R., Sellanes, C., Lou, Y. R., and Raguso, R. A. 2013. Can yeast (*S. cerevisiae*) metabolic volatiles provide polymorphic signaling? *PLoS One* **8**(8), e70219.

Armbruster, S. W. 2002. Can indirect selection and genetic context contribute to trait diversification? A transition-probability study of blossom-colour evolution in two genera. *Journal of Evolutionary Biology* **15**, 468–486.

Ashman, T. L. 2009. Sniffing out patterns of sexual dimorphism in floral scent. *Functional Ecology* **23**, 852–862.

Atwood, J. T. 1985. Pollination of *Paphiopedilum rothschildianum*: brood-site deception. *National Geographic Research* **1**, 247–254.

Ayasse, M., et al. 2000. Evolution of reproductive strategies in the sexually deceptive orchid *Ophrys sphegodes*: how does flower-specific variation of odor signals influence reproductive success? *Evolution* **54**, 1995–2006.

Ayasse, M., Paxton, R. J., and Tengo, J. 2001. Mating behavior and chemical communication in the order Hymenoptera. *Annual Review of Entomology* **46**, 31–78.

Ayasse, M., Schiestl, F. P., Paulus, H. F., Ibarra, F., and Francke, W. 2003. Pollinator attraction in a sexually deceptive orchid by means of unconventional chemicals. *Proceedings of the Royal Society B: Biological Sciences* **270**, 517–522.

Ayasse, M., Scholz, R., Dötterl, S., Francke, W., and Pinheiro, P. M. 2015. Pollinator attraction in the deceptive orchid *Cephalanthera rubra* (Orchidaceae). International Conference about Temperate Orchids, Research and Conservation. Sails for Science Foundation, Samos Greece.

Ayasse, M., Stoekl, J., and Francke, W. 2011. Chemical ecology and pollinator-driven speciation in sexually deceptive orchids. *Phytochemistry* **72**, 1667–1677.

Azuma, H. and Kono, M. 2006. Estragole (4-allylanisole) is the primary compound in volatiles emitted from the male and female cones of *Cycas revoluta*. *Journal of Plant Research* **119**, 671–676.

Backhaus, W., Menzel, R., and Kreissl, S. 1987. Multidimensional-scaling of color similarity in bees. *Biological Cybernetics* **56**, 293–304.

Bailey, S. F., et al. 2007. Empty flowers as a pollination-enhancement strategy. *Evolutionary Ecology Research* **9**, 1245–1262.

Baker, H. G. 1976. "Mistake" pollination as a reproductive system with special reference to the Caricaceae. In: *Tropical Trees: Variation, Breeding and Conservation* (ed. by Burley, J. and Styles, B. T.), pp. 161–166. London: Academic Press.

Baker, H. G. 1984. Some functions of dioecy in seed plants. *The American Naturalist* **124**, 149–158.

Bänziger, H. 1995. Ecological, morphological and taxonomic studies on Thailand's fifth species of Rafflesiaceae: *Rhizanthes zippelii* (Blume) Spach. *Natural History Bulletin of the Siam Society* **43**, 337–365.

Bänziger, H. 1996a. The mesmerizing wart: the pollination strategy of epiphytic lady slipper orchid *Paphiopedilum villosum* (Lindl.) Stein (Orchidaceae). *Botanical Journal of the Linnean Society* **121**, 59–90.

Bänziger, H. 1996b. Pollination of a flowering oddity: *Rhizanthes zippelii* (Blume) Spach (Rafflesiaceae). *Natural History Bulletin of the Siam Society* **44**, 113–142.

Bänziger, H. 2001. Studies on the superlative deceiver: *Rhizanthes* Dumortier (Rafflesiaceae). *Bulletin of the British Ecological Society* **32**, 36–39.

Bänziger, H. 2002. Smart alecks and dumb flies. In: *Proceedings of the 16th World Orchid Congress 1999* (ed. by Clark, J., Elliot, W. M., Tingley, G., and Biro, J.), pp. 165–169. Vancouver: Vancouver Orchid Society.

Bänziger, H. and Pape, T. 2004. Flowers, faeces and cadavers: natural feeding and laying habits of flesh flies in Thailand (Diptera: Sarcophagidae, *Sarcophaga* spp.). *Journal of Natural History* **38**, 1677–1694.

Bänziger, H., Pumikong, S., and Srimuang, K.-O. 2012. The missing link: bee pollination in wild lady slipper orchids *Paphiopedilum thaianum* and *P. niveum* (Orchidaceae) in Thailand. *Mitteilungen der Schweizerischen Entomologischen Gesellschaft* **85**, 1–26.

Bänziger, H., Sun, H. Q., and Luo, Y. B. 2005. Pollination of a slippery lady slipper orchid in south-west China: *Cypripedium guttatum* (Orchidaceae). *Botanical Journal of the Linnean Society* **148**, 251–264.

Barkman, T. J., et al. 2008. Accelerated rates of floral evolution at the upper size limit for flowers. *Current Biology* **18**, 1508–1513.

Barrett, S. C. H., Lloyd, D., and Arroyo, J. 1996. Stylar polymorphisms and the evolution of heterostyly in *Narcissus* (Amaryllidaceae). In: *Floral Biology: Studies on Floral Evolution in Animal-pollinated Plants* (ed. by Lloyd, D. and Barrett, S. C. H.), pp. 339–376. New York: Chapman and Hall.

Barth, F. G. 1991. *Insects and Flowers: The Biology of a Partnership*, 2nd edn. Princeton, NJ: Princeton University Press.

Bateman, R. M., Hollingsworth, P. M., Preston, J., Yi-Bo, L., Pridgeon, A. M., and Chase, M. W. 2003. Molecular phylogenetics and evolution of Orchidinae and selected Habenariinae (Orchidaceae). *Botanical Journal of the Linnean Society* **142**, 1–40.

Bates, H. W. 1862. Contributions to an insect fauna of the Amazon valley (Lepidoptera: Heliconidae). *Transactions of the Linnean Society of London* **23**, 495–566.

Batra, L. R. and Batra, S. W. T. 1985. Floral mimicry induced by mummy-berry fungus exploits host's pollinators as vectors. *Science* **228**, 1011–1013.

Bawa, K. S. 1980a. Evolution of dioecy in flowering plants. *Annual Review of Ecology and Systematics* **11**, 15–39.

Bawa, K. S. 1980b. Mimicry of male by female flowers and intrasexual competition for pollinators in *Jacartia dolichaula* (D. Smith) Woodson (Caricaceae). *Evolution* **34**, 467–474.

Beardsell, D. V., Clements, M. A., Hutchinson, J. F., and Williams, E. G. 1986. Pollination of *Diuris maculata* R. Br. (Orchidaceae) by floral mimicry of the native legumes *Daviesia* spp. and *Pulteanaea scabra* R. Br. *Australian Journal of Botany* **34**, 165–173.

Beath, D. D. N. 1996. Pollination of *Amorphophallus johnsonii* (Araceae) by carrion beetles (*Phaeochrous amplus*) in a Ghanaian rain forest. *Journal of Tropical Ecology* **12**, 409–418.

Becher, P. G., et al. 2012. Yeast, not fruit volatiles mediate *Drosophila melanogaster* attraction, oviposition and development. *Functional Ecology* **26**, 822–828.

Bell, A. K., Roberts, D. L., Hawkins, J. A., Rudall, P. J., Box, M. S., and Bateman, R. M. 2009. Comparative micromorphology of nectariferous and nectarless labellar spurs in selected clades of subtribe Orchidinae (Orchidaceae). *Botanical Journal of the Linnean Society* **160**, 369–387.

Bell, G. 1986. The evolution of empty flowers. *Journal of Theoretical Biology* **118**, 253–258.

Benitez-Vieyra, S., de Ibarra, N. H., Wertlen, A. M., and Cocucci, A. A. 2007. How to look like a mallow: evidence of floral mimicry between Turneraceae and Malvaceae. *Proceedings of the Royal Society B: Biological Sciences* **274**, 2239–2248.

Benitez-Vieyra, S., Medina, A. M., and Cocucci, A. A. 2009. Variable selection patterns on the labellum shape of *Geoblasta pennicillata*, a sexually deceptive orchid. *Journal of Evolutionary Biology* **22**, 2354–2362.

Bergström, G. 1978. Role of volatile chemicals in *Ophrys*–pollinator interactions. In: *Biochemical Aspects of Plant and Animal Coevolution* (ed. by Harborne, G.), pp. 207–230. London: Academic Press.

Bernhardt, P. and Burns, B. P. 1986. Floral mimesis in *Thelymitra nuda* (Orchidaceae). *Plant Systematics and Evolution* **151**, 187–202.

Bierzychudek, P. 1981. *Asclepias, Lantana*, and *Epidendrum*: a floral mimicry complex? *Biotropica* **13**(Suppl.), S54–S58.

Bierzychudek, P. 1982. The demography of Jack-in-the-pulpit, a forest perennial that changes sex. *Ecological Monographs* **52**, 335–351.

Biesmeijer, J. C., Giurfa, M., Koedam, D., Potts, S. G., Joel, D. M., and Dafni, A. 2005. Convergent evolution: floral guides, stingless bee nest entrances, and insectivorous pitchers. *Naturwissenschaften* **92**, 444–450.

Blanco, M. A. and Barboza, G. 2005. Pseudocopulatory pollination in *Lepanthes* (Orchidaceae: Pleurothallidinae) by fungus gnats. *Annals of Botany* **95**, 763–772.

Blanco, M. A., Davies, K. L., Stpiczynska, M., Carlsward, B. S., Ionta, G. M., and Gerlach, G. 2013. Floral elaiophores in *Lockhartia* Hook. (Orchidaceae: Oncidiinae): their distribution, diversity and anatomy. *Annals of Botany* **112**, 1775–1791.

Bohman, B., et al. 2012. Discovery of tetrasubstituted pyrazines as semiochemicals in a sexually deceptive orchid. *Journal of Natural Products* **75**, 1589–1594.

Bohman, B., et al. 2014. Discovery of pyrazines as pollinator sex pheromones and orchid semiochemicals: implications for the evolution of sexual deception. *New Phytologist* **203**, 939–952.

Bolin, J. F., Maass, E., and Musselman, L. J. 2009. Pollination biology of *Hydnora africana* Thunb. (Hydnoraceae) in Namibia: brood-site mimicry with insect imprisonment. *International Journal of Plant Sciences* **170**, 157–163.

Bond, W. J. and Maze, K. E. 1999. Survival costs and reproductive benefits of floral display in a sexually dimorphic dioecious shrub, *Leucadendron xanthoconus*. *Evolutionary Ecology* **13**, 1–18.

Borg-Karlson, A. K., Englund, F. O., and Unelius, C. R. 1994. Dimethyl oligosulphides, major volatiles released from *Sauromatum guttatum* and *Phallus impudicus*. *Phytochemistry* **35**, 321–323.

Borg-Karlson, A.-K. 1990. Chemical and ethological studies of pollination in the genus *Ophrys* (Orchidaceae). *Phytochemistry*, **29**, 1359–1387.

Bower, C. C. 1996. Demonstration of pollinator-mediated reproductive isolation in sexually deceptive species of *Chiloglottis* (Orchidaceae: Caladeniinae). *Australian Journal of Botany* **44**, 15–33.

Bower, C. C., Towle, B., and Bickel, D. 2015. Reproductive success and pollination of the Tuncurry midge orchid (*Genoplesium littorale*) (Orchidaceae) by chloropid flies. *Telopea* **18**, 43–55.

Boyden, T. C. 1980. Floral mimicry by *Epidendrum ibaguense* (Orchidaceae) in Panama. *Evolution* **34**, 135–136.

Boyden, T. C. 1982. The pollination biology of *Calypso bulbosa* var. *americana* (Orchidaceae): initial deception of bumblebee visitors. *Oecologia* **55**, 178–184.

Bradbury, J. W. and Vehrencamp, S. L. 2011. *Principles of Animal Communication*. New York: Sinauer Associates.

Brandenburg, A., Dell'Olivo, A., Bshary, R., and Kuhlemeier, C. 2009. The sweetest thing. Advances in nectar research. *Current Opinion in Plant Biology* **12**, 486–490.

Breitkopf, H., Onstein, R. E., Cafasso, D., Schlueter, P. M., and Cozzolino, S. 2015. Multiple shifts to different pollinators fuelled rapid diversification in sexually deceptive *Ophrys* orchids. *New Phytologist* **207**, 377–389.

Breitkopf, H., Schlueter, P. M., Xu, S., Schiestl, F. P., Cozzolino, S., and Scopece, G. 2013. Pollinator shifts between *Ophrys sphegodes* populations: might adaptation to different pollinators drive population divergence? *Journal of Evolutionary Biology* **26**, 2197–2208.

Bröderbauer, D., Diaz, A., and Weber, A. 2012. Reconstructing the origin and elaboration of insect-trapping inflorescences in the Araceae. *American Journal of Botany* **99**, 1666–1679.

Bröderbauer, D., Weber, A., and Diaz, A. 2013. The design of trapping devices in pollination traps of the genus *Arum* (Araceae) is related to insect type. *Botanical Journal of the Linnean Society* **172**, 385–397.

Brodie, B. S., Babcock, T., Gries, R., Benn, A., and Gries, G. 2016. Acquired smell? Mature female of the common green bottle fly shift semiochemical preferences from feces feeding sites to carrion oviposition sites. *Journal of Chemical Ecology* **42**, 40–50.

Brodie, B., Gries, R., Martins, A., VanLaerhoven, S., and Gries, G. 2014. Bimodal cue complex signifies suitable oviposition sites to gravid females of the common green bottle fly. *Entomologia Experimentalis et Applicata* **153**, 114–127.

Brodie, E. D. and Brodie, E. D. 1980. Differential avoidance of mimetic salamanders by free-ranging birds. *Science* **208**, 181–182.

Brodmann, J., Emer, D., and Ayasse, M. 2012. Pollinator attraction of the wasp-flower *Scrophularia umbrosa* (Scrophulariaceae). *Plant Biology* **14**, 500–505.

Brodmann, J., Twele, R., Francke, W., Holzler, G., Zhang, Q. H., and Ayasse, M. 2008. Orchids mimic green-leaf volatiles to attract prey-hunting wasps for pollination. *Current Biology* **18**, 740–744.

Brodmann, J., Twele, R., Francke, W., Luo, Y. B., Song, X. Q., and Ayasse, M. 2009. Orchid mimics honey bee alarm pheromone in order to attract hornets for pollination. *Current Biology* **19**, 1368–1372.

Bronstein, J. L. 2001. The exploitation of mutualisms. *Ecology Letters* **4**, 277–287.

Brower, L. P. and Brower, J. V. Z. 1972. Parallelism, convergence, divergence, and the new concept of advergence in the evolution of mimicry. *Transactions of the Connecticut Academy of Arts and Science* **4**, 57–67.

Brower, L. P., Hower, A. S., Croze, H. J., Brower, J. V. Z., and Stiles, F. G. 1964. Mimicry—differential advantage of color patterns in the natural environment. *Science* **144**, 183–185.

Brown, A. H. D., Burdon, J. J., and Jarosz, A. M. 1989. Isozyme analysis of plant mating systems. In: *Isozymes in Plant Biology* (ed. by Soltis, D. E. and Soltis, P. S.), pp. 73–86. Portland, OR: Dioscorides Press.

Brown, J. H. and Kodric-Brown, A. 1979. Convergence, competition, and mimicry in a temperate community of hummingbird-pollinated flowers. *Ecology* **60**, 1022–1035.

Brown, J. H. and Kodric-Brown, A. 1981. Reply to Williamson and Black's comment. *Ecology* **62**, 497–498.

Brown, R. L. 1822. An account of a new genus of plants, named *Rafflesia*. *Transactions of the Linnean Society of London* **13**, 210–234.

Brys, R., Jacquemyn, H., and Hermy, M. 2008. Pollination efficiency and reproductive patterns in relation to local plant density, population size, and floral display in the rewarding *Listera ovata* (Orchidaceae). *Botanical Journal of the Linnean Society* **157**, 713–721.

Bultman, T. L. and Leuchtmann, A. 2008. Biology of the *Epichloe–Botanophila* interaction: an intriguing association between fungi and insects. *Fungal Biology Reviews* **22**, 131–138.

Burd, M. 1995. Pollinator behavioural responses to reward size in *Lobelia deckenii*: no escape from pollen limitation of seed set. *Journal of Ecology* **83**, 865–872.

Burger, B. V., Munro, Z. M., and Visser, J. H. 1988. Determination of plant volatiles 0.1. Analysis of the insect-attracting allomone of the parasitic plant *Hydnora africana* using Grob–Habich activated-charcoal traps. *Journal of High Resolution Chromatography and Chromatography Communications* **11**, 496–499.

Cano, L. M., et al. 2013. Major transcriptome reprogramming underlies floral mimicry induced by the rust fungus *Puccinia monoica* in *Boechera stricta*. *PLoS One* **8**(9), e75293.

Carlsson, U. and Elmqvist, T. 1992. Epidemiology of anther-smut disease (*Microbotryum violaceum*) and numeric regulation of populations of *Silene dioica*. *Oecologia* **90**, 509–517.

Carmona-Diaz, G. and Garcia-Franco, J. G. 2009. Reproductive success in the Mexican rewardless *Oncidium cosymbephorum* (Orchidaceae) facilitated by the oil-rewarding *Malpighia glabra* (Malpighiaceae). *Plant Ecology* **203**, 253–261.

Cartar, R. V. 2004. Resource tracking by bumble bees: responses to plant-level differences in quality. *Ecology* **85**, 2764–2771.

Castillo, R. A., Caballero, H., Boege, K., Fornoni, J., and Dominguez, C. A. 2012. How to cheat when you cannot lie? Deceit pollination in *Begonia gracilis*. *Oecologia* **169**, 773–782.

Chan, K. M. A. and Levin, S. A. 2005. Leaky prezygotic isolation and porous genomes: rapid introgression of maternally inherited DNA. *Evolution* **59**, 720–729.

Chang, Y.-Y., et al. 2010. Characterization of the possible roles for B class MADS box genes in regulation of perianth formation in orchid. *Plant Physiology* **152**, 837–853.

Charabidze, D., Depeme, A., Devigne, C., and Hedouin, V. 2015. Do necrophagous blowflies (Diptera: Calliphoridae) lay their eggs in wounds? Experimental data and implications for forensic entomology. *Forensic Science International* **253**, 71–75.

Charlesworth, D. and Charlesworth, B. 1987. Inbreeding depression and its evolutionary consequences. *Annual Review of Ecology and Systematics* **18**, 237–268.

Charnov, E. L. 1976. Optimal foraging, the marginal value theorem. *Theoretical Population Biology* **9**, 129–136.

Chartier, M., Gibernau, M., and Renner, S. S. 2014. The evolution of pollinator–plant interaction types in the Araceae. *Evolution* **68**, 1533–1543.

Chartier, M., Pelozuelo, L., Buatois, B., Bessiere, J.-M., and Gibernau, M. 2013. Geographical variations of odour and pollinators, and test for local adaptation by reciprocal transplant of two European Arum species. *Functional Ecology* **27**, 1367–1381.

Chen, G., et al. 2015. Mimicking livor mortis: a well-known but unsubstantiated color profile in sapromyiophily. *Journal of Chemical Ecology* **41**, 808–815.

Chittka, L. 1992. The color hexagon—a chromaticity diagram based on photoreceptor excitations as a generalized representation of color opponency. *Journal of Comparative Physiology. A, Sensory, Neural, and Behavioral Physiology* **170**, 533–543.

Chittka, L. and Kevan, P. 2005. Flower colour as advertisement. In: *Practical Pollination Biology* (ed. by Dafni, A., Kevan, P. G., and Husband, B. C.), pp. 157–196. Cambridge: Enviroquest.

Chittka, L. and Osorio, D. 2007. Cognitive dimensions of predator responses to imperfect mimicry? *PLoS Biology* **5**, 2754–2758.

Chittka, L., Beier, W., Hertel, H., Steinmann, E., and Menzel, R. 1992. Opponent colour coding is a universal strategy to evaluate the photoreceptor inputs in Hymenoptera. *Journal of Comparative Physiology. A, Sensory, Neural, and Behavioral Physiology* **170**, 545–563.

Chittka, L., Gumbert, A., and Kunze, J. 1997. Foraging dynamics of bumble bees: correlates of movements within and between plant species. *Behavioral Ecology* **8**, 239–249.

Chittka, L., Skorupski, P., and Raine, N. E. 2009. Speed–accuracy tradeoffs in animal decision making. *Trends in Ecology and Evolution* **24**, 400–407.

Chittka, L., Vorobyev, M., Shmida, A., and Menzel, R. 1993. Bee colour vision—the optimal system for the discrimination of flower colours with three spectral photoreceptor types? In: *Sensory Systems of Arthropods* (ed. by Wiese, K., Gribakin, F. G., Popov, A. V., and Renninger, G.), pp. 211–218. Basel, Switzerland: Birkhäuser.

Chou, M.-L., et al. 2013. Global transcriptome analysis and identification of a CONSTANS-like gene family in the orchid *Erycina pusilla*. *Planta* **237**, 1425–1441.

Christiaens, J. F., et al. 2014. The fungal aroma gene ATF1 promotes dispersal of yeast cells through insect vectors. *Cell Reports* **9**, 425–432.

Ciotek, L., Giorgis, P., Benitez-Vieyra, S., and Cocucci, A. A. 2006. First confirmed case of pseudocopulation in terrestrial orchids of South America: pollination of *Geoblasta pennicillata* (Orchidaceae) by *Campsomeris bistrimacula* (Hymenoptera, Scoliidae). *Flora* **201**, 365–369.

Coleman, E. 1927. Pollination of the orchid *Cryptostylis leptochila*. *Victorian Naturalist* **44**, 20–22.

Coleman, E. 1930. Pollination of some west Australian orchids. *Victorian Naturalist* **46**, 203–206.

Cortis, P., Vereecken, N. J., Schiestl, F. P., Lumaga, M. R. B., Scrugli, A., and Cozzolino, S. 2009. Pollinator convergence and the nature of species' boundaries in sympatric Sardinian *Ophrys* (Orchidaceae). *Annals of Botany* **104**, 497–506.

Cozzolino, S., D'Emerico, S., and Widmer, A. 2004. Evidence for reproductive isolate selection in Mediterranean orchids: karyotype differences compensate for the lack of pollinator specificity. *Proceedings of the Royal Society B: Biological Sciences* **271**, S259–S262.

Cozzolino, S. and Widmer, A. 2005. Orchid diversity: an evolutionary consequence of deception? *Trends in Ecology and Evolution* **20**, 487–494.

Cozzolino, S., Schiestl, F. P., Muller, A., De Castro, O., Nardella, A. M., and Widmer, A. 2005. Evidence for pollinator sharing in Mediterranean nectar-mimic orchids: absence of premating barriers? *Proceedings of the Royal Society B: Biological Sciences* **272**, 1271–1278.

Cronquist, A. 1988. *The Evolution and Classification of Flowering Plants*, 2nd edn. New York: New York Botanical Garden.

Cropper, S. C. and Calder, D. M. 1990. The floral biology of *Thelymitra epipactoides* (Orchidaceae), and the implications of pollination by deceit on the survival of this rare orchid. *Plant Systematics and Evolution* **170**, 11–27.

Dafni, A. 1983. Pollination of *Orchis caspia*—a nectarless plant which deceives the pollinators of nectariferous species from other plant families. *Journal of Ecology* **71**, 467–474.

Dafni, A. 1984. Mimicry and deception in pollination. *Annual Review of Ecology and Systematics* **15**, 259–278.

Dafni, A. 1987. Pollination in *Orchis* and related genera: evolution from reward to deception. In: *Orchid Biology: A Review and Perspectives* (ed. by Arditti, J.), pp. 79–104. Ithaca, NY: Comstock Publishing Associates (Cornell University Press).

Dafni, A. and Calder, D. M. 1987. Pollination by deceit and floral mimesis in *Thelymitra antennifera* (Orchidaceae). *Plant Systematics and Evolution* **158**, 11–22.

Dafni, A. and Ivri, Y. 1979. Pollination ecology of, and hybridization between, *Orchis coriophora* L. and *O. collina* Sol. ex Rus. (Orchidaceae) in Israel. *New Phytologist* **83**, 181–187.

Dafni, A. and Ivri, Y. 1981. Floral mimicry between *Orchis israelitica* Baumann and Dafni (Orchidaceae) and *Bellevalia flexuosa* (Liliaceae). *Oecologia* **49**, 229–232.

Dafni, A. and Ivri, Y. 1981. The flower biology of *Cephalanthera longiflora* (Orchidaceae)—pollen imitation and facultative floral mimicry. *Plant Systematics and Evolution* **137**, 229–240.

Dafni, A., Ivri, Y., and Brantjes, N. B. M. 1981. Pollination of *Serapias vomeracea* Briq (Orchidaceae) by imitation of holes for sleeping solitary male bees (Hymenoptera). *Acta Botanica Neerlandica* **30**, 69–73.

Darwin, C. R. 1862. *On the Various Contrivances by which British and Foreign Orchids Are Fertilised by Insects*. London: John Murray.

Darwin, C. R. 1877. *The Various Contrivances by which Orchids Are Fertilised by Insects*, 2nd revised edn. London: John Murray.

Dasmahapatra, K. K., et al. 2012. Butterfly genome reveals promiscuous exchange of mimicry adaptations among species. *Nature* **487**, 94–98.

Davies, K. L., Stpiczynska, M., and Kaminska, M. 2013. Dual deceit in pseudopollen-producing *Maxillaria* s.s. (Orchidaceae: Maxillariinae). *Botanical Journal of the Linnean Society* **173**, 744–763.

Davis, C. C., Endress, P. K., and Baum, D. A. 2008. The evolution of floral gigantism. *Current Opinion in Plant Biology* **11**, 49–57.

Davis, C. C., Latvis, M., Nickrent, D. L., Wurdack, K. J., and Baum, D. A. 2007. Floral gigantism in Rafflesiaceae. *Science* **315**, 1812.

Dawkins, M. and Guilford, T. 1991. The corruption of honest signalling. *Animal Behaviour* **41**, 865–874.

de Ibarra, N. H., Vorobyev, M., and Menzel, R. 2014. Mechanisms, functions and ecology of colour vision in the honeybee. *Journal of Comparative Physiology. A, Neuroethology, Sensory, Neural, and Behavioral Physiology* **200**, 411–433.

de Jager, M. L. and Peakall, R. 2016. Does morphology matter? An explicit assessment of floral morphology in sexual deception. *Functional Ecology* **30**, 537–546.

de Jager, M. L. and Ellis, A. G. 2012. Gender-specific pollinator preference for floral traits. *Functional Ecology* **26**, 1197–1204.

de Jager, M. L. and Ellis, A. G. 2014. Costs of deception and learned resistance in deceptive interactions. *Proceedings of the Royal Society B: Biological Sciences* **281**, 20132861.

de Jager, M. L., Newman, E., Theron, G., Botha, P., Barton, M., and Anderson, B. 2016. Pollinators can prefer rewarding models to mimics: consequences for the assumptions of Batesian floral mimicry. *Plant Systematics and Evolution* **302**, 409–418.

de Jager, M. L. and Peakall, R. 2016. Does morphology matter? An explicit assessment of floral morphology in sexual deception. *Functional Ecology* **30**, 537–546.

Deforge, P. 1995. *Orchids of Britain and Europe*. London: Harper Collins Publishers.

Dekeirsschieter, J., et al. 2009. Cadaveric volatile organic compounds released by decaying pig carcasses (*Sus domesticus* L.) in different biotopes. *Forensic Science International* **189**, 46–53.

Delpino, F. 1868–1875. Ulteriori osservazioni sulla dicogamia nel regno vegetale. *Atti della Societa Italiana di Scienze Naturali Milano*, Vols XI, XII, Pt I (1868/69), Pt II, Fasc. 1 (1870), Fasc. 2 (1875).

Denno, R. F. and Cothran, W. R. 1975. Niche relationships of a guild of necrophagous flies. *Annals of the Entomological Society of America* **68**, 741–754.

Des Marais, D. L. and Rausher, M. D. 2010. Parallel evolution at multiple levels in the origin of hummingbird pollinated flowers in *Ipomea*. *Evolution* **64**, 2044–2054.

Devey, D. S., Bateman, R. M., Fay, M. F., and Hawkins, J. A. 2008. Friends or relatives? Phylogenetics and species delimitation in the controversial European orchid genus *Ophrys*. *Annals of Botany* **101**, 385–402.

Devey, D. S., Bateman, R. M., Fay, M. F., and Hawkins, J. A. 2009. Genetic structure and systematic relationships within the *Ophrys fuciflora* aggregate (Orchidaceae: Orchidinae): high diversity in Kent and a wind-induced discontinuity bisecting the Adriatic. *Annals of Botany* **104**, 483–495.

Diamond, A. R., El Mayas, H., and Boyd, R. S. 2006. *Rudbeckia attriculata* infected with a pollen-mimic fungus in Alabama. *Southeastern Naturalist* **5**, 103–112.

Diaz, A. and Kite, G. C. 2006. Why be a rewarding trap? The evolution of floral rewards in *Arum* (Araceae), a genus characterized by saprophilous pollination systems. *Biological Journal of the Linnean Society* **88**, 257–268.

Dickson, C. R. and Petit, S. 2006. Effect of individual height and labellum colour on the pollination of *Caladenia* (syn. *Arachnorchis*) *behrii* (Orchidaceae) in the northern Adelaide region, South Australia. *Plant Systematics and Evolution* **262**, 65–74.

Dieckmann, U. and Doebeli, M. 1999. On the origin of species by sympatric speciation. *Nature* **400**, 354–357.

Dixon, T. J. 1959. Studies on oviposition behaviour of Syrphidae (Diptera). *Transactions of the Royal Entomological Society of London* **111**, 57–80.

Dobson, H. E. M. and Bergstrom, G. 2000. The ecology and evolution of pollen odors. *Plant Systematics and Evolution* **222**, 63–87.

Donaldson, J. S. 1997. Is there a floral parasite mutualism in cycad pollination? The pollination biology of *Encephalartos villosus* (Zamiaceae). *American Journal of Botany* **84**, 1398–1406.

Dormont, L., Delle-Vedove, R., Bessiere, J. M., Hossaert-Mc Key, M., and Schatz, B. 2010. Rare white-flowered morphs increase the reproductive success of common purple morphs in a food-deceptive orchid. *New Phytologist* **185**, 300–310.

Dötterl, S., Füssel, U., Jürgens, A., and Aas, G. 2005. 1,4-Dimethoxybenzene, a floral scent compound in willows that attracts an oligolectic bee. *Journal of Chemical Ecology* **31**, 2993–2998.

Dötterl, S., Jürgens, A., Seifert, K., Laube, T., Weissbecker, B., and Schutz, S. 2006. Nursery pollination by a moth in *Silene latifolia*: the role of odours in eliciting antennal and behavioural responses. *New Phytologist* **169**, 707–718.

Dötterl, S., Jürgens, A., Wolfe, L., and Biere, A. 2009. Disease status and population origin effects on floral scent: potential consequences for oviposition and fruit predation in a complex interaction between a plant, fungus, and noctuid moth. *Journal of Chemical Ecology* **35**, 307–319.

Dötterl, S., Milchreit, K., and Schaffler, I. 2011. Behavioural plasticity and sex differences in host finding of a specialized bee species. *Journal of Comparative Physiology. A, Neuroethology, Sensory, Neural, and Behavioral Physiology* **197**, 1119–1126.

Dressler, R. L. 1981. *The Orchids: Natural History and Classification*. Cambridge, MA: Harvard University Press.

Duffy, K. J. and Johnson, S. D. 2015. Staminal hairs enhance fecundity in the pollen-rewarding self-incompatible lily *Bulbine abyssinica*. *Botanical Journal of the Linnean Society* **177**, 481–490.

Dukas, R. 1987. Foraging behavior of three bee species in a natural mimicry system: female flowers which mimic male flowers in *Ecballium elaterium* (Cucurbitaceae). *Oecologia* **74**, 256–263.

Dukas, R. and Real, L. A. 1991. Learning foraging tasks by bees—a comparison between social and solitary species. *Animal Behaviour* **42**, 269–276.

Dyer, A. G. and Murphy, A. H. 2009. Honeybees choose "incorrect" colors that are similar to target flowers in preference to novel colors. *Israel Journal of Plant Sciences* **57**, 203–210.

Eberhard, W. G. 1977. Aggressive chemical mimicry by a bolus spider. *Science* **198**, 1173–1175.

Edens-Meier, R., Raguso, R. A., Westhus, E., and Bernhardt, P. 2014. Floral fraudulence: do blue *Thelymitra* species (Orchidaceae) mimic *Orthrosanthus laxus* (Iridaceae). *Telopea* **17**, 15–28.

Edmunds, M. 2000. Why are there good and poor mimics? *Biological Journal of the Linnean Society* **70**, 459–466.

Ellis, A. G. and Johnson, S. D. 2010. Floral mimicry enhances pollen export: the evolution of pollination by sexual deceit outside of the Orchidaceae. *The American Naturalist* **176**, E143–E151.

Endara, L., Grimaldi, D., and Roy, B. A. 2010. Lord of the flies: pollination of *Dracula* orchids. *Lankesteriana* **10**, 1–11.

Endler, J. A. 1981. An overview of the relationships between mimicry and crypsis. *Biological Journal of the Linnean Society* **16**, 25–31.

Endler, J. A. and Basolo, A. L. 1998. Sensory ecology, receiver biases and sexual selection. *Trends in Ecology and Evolution* **13**, 415–420.

Faegri, K. and van der Pijl, L. 1979. *The Principles of Pollination Ecology*. Oxford: Pergamon.

Faldt, J., Jonsell, M., Nordlander, G., and Borg-Karlson, A. K. 1999. Volatiles of bracket fungi *Fomitopsis pinicola* and *Fomes fomentarius* and their functions as insect attractants. *Journal of Chemical Ecology* **25**, 567–590.

Fatouros, N. E., Huigens, M. E., van Loon, J. J. A., Dicke, M., and Hilker, M. 2005. Chemical communication—butterfly anti-aphrodisiac lures parasitic wasps. *Nature* **433**, 704–704.

Feinsinger, P. 1987. Effects of plant species on each other's pollination: is community structure influenced? *Trends in Ecology and Evolution* **2**, 123–126.

Fenster, C. B., Armbruster, W. S., Wilson, P., Dudash, M. R., and Thomson, J. D. 2004. Pollination syndromes and floral specialization. *Annual Review of Ecology Evolution and Systematics* **35**, 375–403.

Ferguson, L. C., Maroja, L., and Jiggins, C. D. 2011. Convergent, modular expression of ebony and tan in the mimetic wing patterns of *Heliconius* butterflies. *Development Genes and Evolution* **221**, 297–308.

Feulner, M., Lauerer, M., and Dötterl, S. 2015. Es stinkt! Komponenten im Blütenduft von *Orchidantha fimbriata*. *Der Palmengarten* **79**, 52–58.

Fisher, R. A. 1930. *The Genetical Theory of Natural Selection*. Oxford: Oxford University Press.

Forbes, P. 2009. *Dazzled and Deceived: Mimicry and Camouflage*. New Haven, CT: Yale University Press.

Ford, E. B. 1975. *Ecological Genetics*, 4th edn. London: Chapman and Hall.

Fordham, F. 1946. Pollination of *Calochilus campestris*. *Victorian Naturalist* **62**, 199–201.

Forrest, J. and Thomson, J. D. 2009. Background complexity affects colour preference in bumblebees. *Naturwissenschaften* **96**, 921–925.

Fritz, A.L. 1990. Deceit pollination of *Orchid spitzelii* (Orchidaceae) on the island of Gotland in the Baltic: a suboptimal system. *Nordic Journal of Botany* **9**, 577–587.

Fritz, A. L. and Nilsson, L. A. 1994. How pollinator-mediated mating varies with population-size in plants. *Oecologia* **100**, 451–462.

Fuhro, D., De Araujo, A. M., and Irgang, B. E. 2010. Are there evidences of a complex mimicry system among *Asclepias curassavica* (Apocynaceae), *Epidendrum fulgens* (Orchidaceae), and *Lantana camara* (Verbenaceae) in southern Brazil? *Revista Brasileira de Botanica* **33**, 589–598.

Galen, C. and Plowright, R. C. 1985. The effects of nectar level and flower development on pollen carry-over in inflorescences of fireweed (*Epilobium angustifolium*) (Onagraceae). *Canadian Journal of Botany* **63**, 488–491.

Galizia, C. G., et al. 2005. Relationship of visual and olfactory signal parameters in a food-deceptive flower mimicry system. *Behavioral Ecology* **16**, 159–168.

Galliot, C., Stuurman, J., and Kuhlemeier, C. 2006. The genetic dissection of floral pollination syndromes. *Current Opinion in Plant Biology* **9**, 78–82.

Gamberale-Stille, G., Balogh, A. C. V., Tullberg, B. S., and Leimar, O. 2012. Feature saltation and the evolution of mimicry. *Evolution* **66**, 807–817.

Gaskett, A. C. 2011. Orchid pollination by sexual deception: pollinator perspectives. *Biological Reviews* **86**, 33–75.

Gaskett, A. C. 2012. Floral shape mimicry and variation in sexually deceptive orchids with a shared pollinator. *Biological Journal of the Linnean Society* **106**, 469–481.

Gaskett, A. C. and Herberstein, M. E. 2010. Colour mimicry and sexual deception by tongue orchids (*Cryptostylis*). *Naturwissenschaften* **97**, 97–102.

Gaskett, A. C., Winnick, C. G., and Herberstein, M. E. 2008. Orchid sexual deceit provokes ejaculation. *The American Naturalist* **171**, E206–E212.

Gebauer, G. and Meyer, M. 2003. N-15 and C-13 natural abundance of autotrophic and mycoheterotrophic orchids provides insight into nitrogen and carbon gain from fungal association. *New Phytologist* **160**, 209–223.

Gigord, L. D. B., Macnair, M. M., and Smithson, A. 2001. Negative frequency dependent selection maintains a dramatic flower color polymorphism in the rewardless orchid *Dactylorhiza sambucina* (L.) Soò. *Proceedings of the National Academy of Sciences of the United States of America* **98**, 6253–6255.

Gigord, L. D. B., Macnair, M. R., Stritesky, M., and Smithson, A. 2002. The potential for floral mimicry in rewardless orchids: an experimental study. *Proceedings of the Royal Society B: Biological Sciences* **269**, 1389–1395.

Gilbert, F. 2005. The evolution of imperfect mimicry. In: *Insect Evolutionary Ecology* (ed. by Fellowes, M. D. E., Holloway, G. J., and Rolff, J.), pp. 231–288. Wallingford: CABI.

Gilbert, F. S., Haines, N., and Dickson, K. 1991. Empty flowers. *Functional Ecology* **5**, 29–39.

Gill, D. E. 1989. Fruiting failure, pollinator inefficiency, and speciation in orchids. In: *Speciation and its Consequences* (ed. by Otte, D. and Endler, J. A.), pp. 458–481. Sunderland, MA: Sinauer.

Giurfa, M., Nunez, J., Chittka, L., and Menzel, R. 1995. Color preferences of flower-naive honeybees. *Journal of Comparative Physiology. A, Sensory, Neural, and Behavioral Physiology* **177**, 247–259.

Giurfa, M. and Sandoz, J.-C. 2012. Invertebrate learning and memory: fifty years of olfactory conditioning of the proboscis extension response in honeybees. *Learning and Memory* **19**, 54–66.

Giurfa, M., Vorobyev, M., Brandt, R., Posner, B., and Menzel, R. 1997. Discrimination of coloured stimuli by honeybees: alternative use of achromatic and chromatic signals. *Journal of Comparative Physiology. A, Sensory, Neural, and Behavioral Physiology* **180**, 235–243.

Givnish, T. J., et al. 2015. Orchid phylogenomics and multiple drivers of their extraordinary diversification. *Proceedings of the Royal Society B: Biological Sciences* **282**, doi: 10.1098/rspb.2015.1553.

Godfery, M. J. 1925. The fertilisation of *Ophrys speculum, O. lutea*, and *O. fusca. Journal of Botany* **63**, 33–40.

Gögler, J., Twele, R., Francke, W., and Ayasse, M. 2011. Two phylogenetically distinct species of sexually deceptive orchids mimic the sex pheromone of their single common pollinator, the cuckoo bumblebee *Bombus vestalis. Chemoecology* **21**, 243–252.

Golubov, J., Mandujano, M. C., Montana, C., Lopez-Portillo, J., and Eguiarte, L. E. 2004. The demographic costs of nectar production in the desert perennial *Prosopis glandulosa* (Mimosoideae): a modular approach. *Plant Ecology* **170**, 267–275.

Goodrich, K. R. 2012. Floral scent in Annonaceae. *Botanical Journal of the Linnean Society* **169**, 262–279.

Goodrich, K. R., Zjhra, M. L., Ley, C. A., and Raguso, R. A. 2006. When flowers smell fermented: the chemistry and ontogeny of yeasty floral scent in pawpaw (*Asimina triloba*: Annonaceae). *International Journal of Plant Sciences* **167**, 33–46.

Gosden, T. P. and Svensson, E. I. 2009. Density-dependent male mating harassment, female resistance, and male mimicry. *The American Naturalist* **173**, 709–721.

Gottsberger, G. 2012. How diverse are Annonaceae with regard to pollination? *Botanical Journal of the Linnean Society* **169**, 245–261.

Gould, S. J. and Vrba, E. S. 1982. Exaptation—a missing term in the science of form. *Paleobiology* **8**, 4–15.

Goyret, J., Pfaff, M., Raguso, R. A., and Kelber, A. 2008. Why do *Manduca sexta* feed from white flowers? Innate and learnt colour preferences in a hawkmoth. *Naturwissenschaften* **95**, 569–576.

Grant, K. A. 1966. A hypothesis concerning prevalence of red coloration in California hummingbird flowers. *The American Naturalist* **100**, 85–97.

Grant, P. R., Grant, B. R., Markert, J. A., Keller, L. F., and Petren, K. 2004. Convergent evolution of Darwin's finches caused by introgressive hybridization and selection. *Evolution* **58**, 1588–1599.

Grant, V. 1949. Pollination systems as isolating mechanisms in angiosperms. *Evolution* **3**, 82–97.

Grant, V. 1994. Modes and origins of mechanical and ethological isolation in angiosperms. *Proceedings of the National Academy of Sciences of the United States of America* **91**, 3–10.

Grant, V. and Grant, K. A. 1965. *Flower Pollination in the Phlox Family*. New York: Columbia University Press.

Greene, H. W. and McDiarmid, R. W. 1981. Coral snake mimicry: does it occur? *Science* **213**, 1207–1212.

Grimaldi, D. and Engel, M. S. 2005. *Evolution of the Insects*. New York: Cambridge University Press.

Grison-Pige, L., Bessiere, J. M., Turlings, T. C. J., Kjellberg, F., Roy, J., and Hossaert-McKey, M. M. 2001. Limited intersex mimicry of floral odour in *Ficus carica*. *Functional Ecology* **15**, 551–558.

Guerrieri, F., Schubert, M., Sandoz, J. C., and Giurfa, M. 2005. Perceptual and neural olfactory similarity in honeybees. *PLoS Biology* **3**, 718–732.

Gumbert, A. 2000. Color choices by bumble bees (*Bombus terrestris*): innate preferences and generalization after learning. *Behavioral Ecology and Sociobiology* **48**, 36–43.

Gumbert, A. and Kunze, J. 2001. Colour similarity to rewarding model affects pollination in a food deceptive orchid, *Orchis boryi*. *Biological Journal of the Linnean Society* **72**, 419–433.

Harder, L. D. 2000. Pollen dispersal and the floral diversity of monocotyledons. In: *Monocots: Systematics and Evolution* (ed. by Wilson, K. L. and Morrison, D. A.), pp. 243–257. Melbourne: CSIRO.

Harder, L. D. and Barrett, S. C. H. 1992. The energy cost of bee pollination for *Pontederia cordata* (Pontederiaceae). *Functional Ecology* **6**, 226–233.

Harder, L. D. and Barrett, S. C. H. 1995. Mating cost of large floral displays in hermaphrodite plants. *Nature* **373**, 512–515.

Harder, L. D. and Johnson, S. D. 2008. Function and evolution of aggregated pollen in angiosperms. *International Journal of Plant Sciences* **169**, 59–78.

Harder, L. D. and Thomson, J. D. 1989. Evolutionary options for maximizing pollen dispersal of animal-pollinated plants. *The American Naturalist* **133**, 323–344.

Harder, L. D. and Wilson, W. G. 1994. Floral evolution and male reproductive success: optimal dispensing schedules for pollen dispersal by animal-pollinated plants. *Evolutionary Ecology* **8**, 542–559.

Harder, L. D., Williams, N. M., Jordan, C. Y., and Nelson, W. A. 2001. The effects of floral design and display on pollinator economics and pollen dispersal. In: *Cognitive Ecology of Pollination* (ed. by Chittka, L. and Thomson, J. D.), pp. 297–317. Cambridge: Cambridge University Press.

Harlin, C. and Harlin, M. 2003. Towards a historization of aposematism. *Evolutionary Ecology* **17**, 197–212.

Harvey, P. H. and Paxton, R. J. 1981. The evolution of aposematic coloration. *Oikos* **37**, 391–393.

Heiduk, A., et al. 2015. Deceptive *Ceropegia dolichophylla* fools its kleptoparasitic fly pollinators with exceptional floral scent. *Frontiers in Ecology and Evolution* **3**, 66, doi: 10.3389/fevo.2015.00066.

Heil, M. 2011. Nectar: generation, regulation, and ecological functions. *Trends in Plant Science* **16**, 191–200.

Heinrich, B. 1975. Bee flowers: a hypothesis on flower variety and blooming times. *Evolution* **29**, 325–334.

Heinrich, B. 1979. Resource heterogeneity and patterns of movement in foraging bumblebees. *Oecologia* **40**, 235–245.

Hemborg, A. M. and Bond, W. J. 2005. Different rewards in female and male flowers can explain the evolution of sexual dimorphism in plants. *Biological Journal of the Linnean Society* **85**, 97–109.

Hermann, K., Klahre, U., Moser, M., Sheehan, H., Mandel, T., and Kuhlemeier, C. 2013. Tight genetic linkage of prezygotic barrier loci creates a multifunctional speciation island in *Petunia*. *Current Biology* **23**, 873–877.

Hermann, K., Klahre, U., Venail, J., Brandenburg, A., and Kuhlemeier, C. 2015. The genetics of reproductive organ morphology in two *Petunia* species with contrasting pollination syndromes. *Planta* **241**, 1241–1254.

Heuschen, B., Gumbert, A., and Lunau, K. 2005. A generalised mimicry system involving angiosperm flower colour, pollen and bumblebees' innate colour preferences. *Plant Systematics and Evolution* **252**, 121–137.

Hines, H. M., et al. 2011. Wing patterning gene redefines the mimetic history of *Heliconius* butterflies. *Proceedings of the National Academy of Sciences of the United States of America* **108**, 19666–19671.

Hobbhahn, N., Johnson, S. D., Bytebier, B., Yeung, E. C., and Harder, L. D. 2013. The evolution of floral nectaries in *Disa* (Orchidaceae: Disinae): recapitulation or diversifying innovation? *Annals of Botany* **112**, 1303–1319.

Hodges, S. A. 1995. The influence of nectar production on hawkmoth behavior, self pollination, and seed production in *Mirabilis multiflora* (Nyctaginaceae). *American Journal of Botany* **82**, 197–204.

Hodges, S. A. and Derieg, N. J. 2009. Adaptive radiations: from field to genomic studies. *Proceedings of the National Academy of Sciences of the United States of America* **106**, 9947–9954.

Hopper, S. D. and Brown, A. P. 2007. A revision of Australia's hammer orchids (*Drakaea*: Orchidaceae), with some field data on species-specific sexually deceived wasp pollinators. *Australian Systematic Botany* **20**, 252–285.

Howell, A. D. and Alarcon, R. 2007. *Osmia* bees (Hymenoptera: Megachilidae) can detect nectar-rewarding flowers using olfactory cues. *Animal Behaviour* **74**, 199–205.

Hsiao, Y.-Y., et al. 2011. Research on orchid biology and biotechnology. *Plant and Cell Physiology* **52**, 1467–1486.

Hsieh, M.-H., et al. 2013. Optimizing virus-induced gene silencing efficiency with *Cymbidium* mosaic virus in *Phalaenopsis* flower. *Plant Science* **201**, 25–41.

Hsu, H. F., et al. 2015. Model for perianth formation in orchids. *Nature Plants* **1**, art. no. 15046.

Hsu, H.-F. and Yang, C.-H. 2002. An orchid (*Oncidium* Gower Ramsey) *AP3*-like MADS gene regulates floral formation and initiation. *Plant and Cell Physiology* **43**, 1198–1209.

Husband, B. C. and Schemske, D. W. 1996. Evolution of the magnitude and timing of inbreeding depression in plants. *Evolution* **50**, 54–70.

Inda, L. A., Pimentel, M., and Chase, M. W. 2012. Phylogenetics of tribe Orchideae (Orchidaceae: Orchidoideae) based on combined DNA matrices: inferences regarding timing of diversification and evolution of pollination syndromes. *Annals of Botany* **110**, 71–90.

Indsto, J. O., Weston, P. H., Clements, M. A., Dyer, A. G., Batley, M., and Whelan, R. J. 2006. Pollination of *Diuris maculata* (Orchidaceae) by male *Trichocolletes venustus* bees. *Australian Journal of Botany* **54**, 669–679.

Internicola, A. I. and Harder, L. D. 2012. Bumble-bee learning selects for both early and long flowering in food-deceptive plants. *Proceedings of the Royal Society B: Biological Sciences* **279**, 1538–1543.

Internicola, A. I., Bernasconi, G., and Gigord, L. D. B. 2008. Should food-deceptive species flower before or after rewarding species? An experimental test of pollinator visitation behaviour under contrasting phenologies. *Journal of Evolutionary Biology* **21**, 1358–1365.

Internicola, A. I., Juillet, N., Smithson, A., and Gigord, L. D. B. 2006. Experimental investigation of the effect of spatial aggregation on reproductive success in a rewardless orchid. *Oecologia* **150**, 435–441.

Irwin, R. E. 2003. Impact of nectar robbing on estimates of pollen flow: conceptual predictions and empirical outcomes. *Ecology* **84**, 485–495.

Irwin, R. E. and Brody, A. K. 1998. Nectar robbing in *Ipomopsis aggregata*: effects on pollinator behavior and plant fitness. *Oecologia* **116**, 519–527.

Irwin, R. E., Brody, A. K., and Waser, N. M. 2001. The impact of floral larceny on individuals, populations, and communities. *Oecologia* **129**, 161–168.

Irwin, R. E., Bronstein, J. L., Manson, J. S., and Richardson, L. 2010. Nectar robbing: ecological and evolutionary perspectives. *Annual Review of Ecology, Evolution, and Systematics*, **41**, 271–292.

Ito, K. and Seymour, R. S. 2005. Expression of uncoupling protein and alternative oxidase depends on lipid or carbohydrate substrates in thermogenic plants. *Biology Letters* **1**, 427–430.

Ivri, Y. and Dafni, A. 1977. The pollination ecology of *Epipactis consimilis* Don (Orchidaceae) in Israel. *New Phytologist* **79**, 173–177.

Jacquemyn, H. and Brys, R. 2010. Temporal and spatial variation in flower and fruit production in a food-deceptive orchid: a five-year study. *Plant Biology* **12**, 145–153.

Jersáková, J., et al. 2010. Absence of pollinator-mediated premating barriers in mixed-ploidy populations of *Gymnadenia conopsea* s.l. (Orchidaceae). *Evolutionary Ecology* **24**, 1199–1218.

Jersáková, J., et al. 2016. Does the globe orchid dupe its pollinators through generalized food deception or mimicry? *Botanical Journal of the Linnean Society* **180**, 269–294.

Jersáková, J. and Johnson, S. D. 2006. Lack of floral nectar reduces self-pollination in a fly-pollinated orchid. *Oecologia* **147**, 60–68.

Jersáková, J., Johnson, S. D., and Kindlmann, P. 2006. Mechanisms and evolution of deceptive pollination in orchids. *Biological Reviews* **81**, 219–235.

Jersáková, J., Johnson, S. D., Kindlmann, P., and Pupin, A. C. 2008. Effect of nectar supplementation on male and female components of pollination success in the deceptive orchid *Dactylorhiza sambucina*. *Acta Oecologica–International Journal of Ecology* **33**, 300–306.

Jersáková, J., Jürgens, A., Smilauer, P., and Johnson, S. D. 2012. The evolution of floral mimicry: identifying traits that visually attract pollinators. *Functional Ecology* **26**, 1381–1389.

Jersáková, J., Kindlmann, P., and Renner, S. S. 2006b. Is the colour dimorphism in *Dactylorhiza sambucina* maintained by differential seed viability instead of frequency-dependent selection? *Folia Geobotanica* **41**, 61–76.

Jiggins, C. D., Naisbit, R. E., Coe, R. L., and Mallet, J. 2001. Reproductive isolation caused by colour pattern mimicry. *Nature* **411**, 302–305.

Jin, X. H., Ren, Z. X., Xu, S. Z., Wang, H., Li, D. Z., and Li, Z. Y. 2014. The evolution of floral deception in *Epipactis veratrifolia* (Orchidaceae): from indirect defense to pollination. *BMC Plant Biology* **14**, 63.

Johnson, S. D. 1994. Evidence for Batesian mimicry in a butterfly-pollinated orchid. *Biological Journal of the Linnean Society* **53**, 91–104.

Johnson, S. D. 2000. Batesian mimicry in the non-rewarding orchid *Disa pulchra*, and its consequences for pollinator behaviour. *Biological Journal of the Linnean Society* **71**, 119–132.

Johnson, S. D. 2010. The pollination niche and its role in the diversification and maintenance of the southern African flora. *Philosophical Transactions of the Royal Society B: Biological Sciences* **365**, 499–516.

Johnson, S. D. and Bond, W. J. 1997. Evidence for widespread pollen limitation of fruiting success in Cape wildflowers. *Oecologia* **109**, 530–534.

Johnson, S. D. and Dafni, A. 1998. Response of bee-flies to the shape and pattern of model flowers: implications for floral evolution in a Mediterranean herb. *Functional Ecology* **12**, 289–297.

Johnson, S. D. and Jürgens, A. 2010. Convergent evolution of carrion and faecal scent mimicry in fly-pollinated angiosperm flowers and a stinkhorn fungus. *South African Journal of Botany* **76**, 796–807.

Johnson, S. D. and Morita, S. 2006. Lying to Pinocchio: floral deception in an orchid pollinated by long-proboscid flies. *Botanical Journal of the Linnean Society* **152**, 271–278.

Johnson, S. D. and Nilsson, L. A. 1999. Pollen carryover, geitonogamy and the evolution of deceptive pollination systems in orchids. *Ecology* **80**, 2607–2619.

Johnson, S. D. and Raguso, R. 2016. The long-tongued hawkmoth pollinator niche for native and invasive plants in Africa. *Annals of Botany* **117**, 25–36.

Johnson, S. D. and Steiner, K. E. 1994. Pollination by megachilid bees and determinants of fruit-set in the Cape orchid *Disa tenuifolia*. *Nordic Journal of Botany* **14**, 481–485.

Johnson, S. D., Alexandersson, R., and Linder, H. P. 2003a. Experimental and phylogenetic evidence for floral mimicry in a guild of fly-pollinated plants. *Biological Journal of the Linnean Society* **80**, 289–304.

Johnson, S. D., Hobbhahn, N., and Bytebier, B. 2013. Ancestral deceit and labile evolution of nectar production in the African orchid genus *Disa* . *Biology Letters* **9**, 20130500.

Johnson, S. D., Hollens, H., and Kuhlmann, M. 2012. Competition versus facilitation: conspecific effects on pollinator visitation and seed set in the iris *Lapeirousia oreogena*. *Oikos* **121**, 545–550.

Johnson, S. D., Linder, H. P., and Steiner, K. E. 1998. Phylogeny and radiation of pollination systems in *Disa* (Orchidaceae). *American Journal of Botany* **85**, 402–411.

Johnson, S. D., Peter, C. I., and Ågren, J. 2004. The effects of nectar addition on pollen removal and geitonogamy in the non-rewarding orchid *Anacamptis morio*. *Proceedings of the Royal Society B: Biological Sciences* **271**, 803–809.

Johnson, S. D., Peter, C. I., Nilsson, L. A., and Ågren, J. 2003b. Pollination success in a deceptive orchid is enhanced by co-occurring rewarding magnet plants. *Ecology* **84**, 2919–2927.

Johnson, S. D., Steiner, K. E., and Kaiser, R. 2005. Deceptive pollination in two subspecies of *Disa spathulata* (Orchidaceae) differing in morphology and floral fragrance. *Plant Systematics and Evolution* **255**, 87–98.

Johnson, S. D., Torninger, E., and Ågren, J. 2009. Relationships between population size and pollen fates in a moth-pollinated orchid. *Biology Letters* **5**, 282–285.

Joron, M. and Mallet, J. L. B. 1998. Diversity in mimicry: paradox or paradigm. *Trends in Ecology and Evolution* **13**, 461–466.

Juillet, N. and Scopece, G. 2010. Does floral trait variability enhance reproductive success in deceptive orchids? *Perspectives in Plant Ecology, Evolution, and Systematics* **12**, 317–322.

Juillet, N., Gonzalez, M. A., Page, P. A., and Gigord, L. D. B. 2007. Pollination of the European food-deceptive *Traunsteinera globosa* (Orchidaceae): the importance of nectar-producing neighbouring plants. *Plant Systematics and Evolution* **265**, 123–129.

Juillet, N., Salzmann, C. C., and Scopece, G. 2011. Does facilitating pollinator learning impede deceptive orchid attractiveness? A multi-approach test of avoidance learning. *Plant Biology* **13**, 570–575.

Jürgens, A. and Shuttleworth, A. 2015. Carrion and dung mimicry in plants. In: *Carrion Ecology, Evolution and Their Applications* (ed. by Benbow, M. E., Tomberlin, J. K., and Tarone, A. M.), pp. 361–386. Boca Raton, FL: CRC Press.

Jürgens, A., Dötterl, S., and Meve, U. 2006. The chemical nature of fetid floral odours in stapeliads (Apocynaceae–Asclepiadoideae–Ceropegieae). *New Phytologist* **172**, 452–468.

Jürgens, A., Wee, S.-L., Shuttleworth, A., and Johnson, S. D. 2013. Chemical mimicry of insect oviposition sites: a global analysis of convergence in angiosperms. *Ecology Letters* **16**, 1157–1167.

Kaiser, R. 1993. *The Scent of Orchids. Olfactory and Chemical Investigations*, Elsevier Amsterdam.

Kaiser, R. 2006. Flowers and fungi use scents to mimic each other. *Science* **311**, 806–807.

Kalinova, B., Podskalska, H., Ruzicka, J., and Hoskovec, M. 2009. Irresistible bouquet of death-how are burying beetles (Coleoptera: Silphidae: Nicrophorus) attracted by carcasses. *Naturwissenschaften* **96**, 889–899.

Kay, A. O. N. 1978. The role of preferential and assortative pollination in the maintenance of flower colour polymorphisms. In: *The Pollination of Flowers by Insects* (ed. by Richards, A. J.), pp. 175–190. London: Academic Press.

Kay, Q. O. N. 1982. Intraspecific discrimination by pollinators and its role in evolution. In: *Pollination and Evolution* (ed. by Armstrong, J. A., Powell, J. M., and Richards, A. J.), pp. 9–28. Sydney: Royal Botanic Gardens.

Kelber, A. 1996. Colour learning in the hawkmoth *Macroglossum stellatarum*. *Journal of Experimental Biology* **199**, 1127–1131.

Kelber, A., Vorobyev, M., and Osorio, D. 2003. Animal colour vision—behavioural tests and physiological concepts. *Biological Reviews* **78**, 81–118.

Kelly, M. M. and Gaskett, C. 2014. UV reflectance but no evidence for colour mimicry in a putative brood-deceptive orchid *Corybas cheesemanii*. *Current Zoology* **60**, 104–113.

Kessler, D., Kallenbach, M., Diezel, C., Rothe, E., Murdock, M., and Baldwin, I. T. 2015. How scent and nectar influence floral antagonists and mutualists. *eLife* **4**, e07641.

Kikuchi, D. W. and Pfennig, D. W. 2010. High-model abundance may permit the gradual evolution of Batesian mimicry: an experimental test. *Proceedings of the Royal Society B: Biological Sciences* **277**, 1041–1048.

Kikuchi, D. W. and Pfennig, D. W. 2013. Imperfect mimicry and the limits of natural selection. *Quarterly Review of Biology* **88**, 297–315.

Kikuchi, D. W., Seymoure, B. M., and Pfennig, D. W. 2014. Mimicry's palette: widespread use of conserved pigments in the aposematic signals of snakes. *Evolution and Development* 16, 61–67.

Kitchener, F. E. 1867. Darwin Correspondence Project, letter no. 5674. Available at: http://www.darwinproject.ac.uk/DCP-LETT-5674 (accessed April 8, 2016).

Kite, G. C. 1995. The floral odour of *Arum maculatum*. *Biochemical Systematics and Ecology* **23**, 343–354.

Kite, G. C. and Hetterscheid, W. L. A. 1997. Inflorescence odours of *Amorphophallus* and *Pseudodracontium* (Araceae). *Phytochemistry* **46**, 71–75.

Kite, G. C., et al. 1998. Inflorescence odours and pollinators of *Arum* and *Amorphophallus* (Araceae). In: *Reproductive Biology* (ed. by Owens, S. J. and Rudall, P. J.), pp. 295–315. Kew: Royal Botanical Gardens.

Kjellsson, G., Rasmussen, F. N., and Dupuy, D. 1985. Pollination of *Dendrobium infundibulum, Cymbidium insigne* (Orchidaceae) and *Rhododendron lyi* (Ericaceae) by *Bombus eximius* (Apidae) in Thailand: a possible case of floral mimicry. *Journal of Tropical Ecology* **1**, 289–302.

Klahre, U., et al. 2011. Pollinator choice in *Petunia* depends on two major genetic loci for floral scent production. *Current Biology* **21**, 730–739.

Klinkhamer, P. G. L. and de Jong, T. J. 1993. Attractiveness to pollinators: a plant's dilemma. *Oikos* **66**, 180–183.

Klinkhamer, P. G. L., de Jong, T. J., and Linnebank, L. A. 2001. Small-scale spatial patterns determine ecological relationships: an experimental example using nectar production rates. *Ecology Letters* **4**, 559–567.

Klinkhamer, P. G. L. and van der Lugt, P. P. 2004. Pollinator service only depends on nectar production rates in sparse populations. *Oecologia* **140**, 491–494.

Knauer, A. C. and Schiestl, F. P. 2015. Bees use honest floral signals as indicators of reward when visiting flowers. *Ecology Letters* **18**, 135–143.

Knight, T. M., et al. 2005. Pollen limitation of plant reproduction: pattern and process. *Annual Review of Ecology Evolution and Systematics* **36**, 467–497.

Knudsen, J. T. and Tollsten, L. 1995. Floral scent in bat-pollinated plants: a case of convergent evolution. *Botanical Journal of the Linnean Society* **119**, 45–57.

Koivisto, A. M., Vallius, E., and Salonen, V. 2002. Pollination and reproductive success of two colour variants of a deceptive orchid, *Dactylorhiza maculata* (Orchidaceae). *Nordic Journal of Botany* **22**, 53–58.

Kono, M. and Tobe, H. 2007. Is *Cycas revoluta* (Cycadeae) wind- or insect-pollinated? *American Journal of Botany* **94**, 847–855.

Koopowitz, H. and Marchant, T. A. 1998. Postpollination nectar reabsorption in the African epiphyte *Aerangis verdickii* (Orchidaceae). *American Journal of Botany* **85**, 508–512.

Kronforst, M. R. and Papa, R. 2015. The functional basis of wing patterning in *Heliconius* butterflies: the molecules behind mimicry. *Genetics* **200**, 1–19.

Kropf, M. and Renner, S. S. 2008. Pollinator-mediated selfing in two deceptive orchids and a review of pollinium tracking studies addressing geitonogamy. *Oecologia* **155**, 497–508.

Kullenberg, B. 1961. Studies in *Ophrys* pollination. *Zoologiska bidrag från Uppsala*, **34**, 1–340.

Kunte, K., et al. 2014. Doublesex is a mimicry supergene. *Nature* **507**, 229–232.

Kunze, J. and Gumbert, A. 2001. The combined effect of color and odor on flower choice behavior of bumble bees in flower mimicry systems. *Behavioral Ecology* **12**, 447–456.

Kuusela, S. and Hanski, I. 1982. The structure of carrion fly communities—the size and the type of carrion. *Holarctic Ecology* **5**, 337–348.

Lammi, A. and Kuitunen, M. 1995. Deceptive pollination of *Dactylorhiza incarnata*: an experimental test of the magnet species hypothesis. *Oecologica* **101**, 500–503.

Lande, R. and Schemske, D. W. 1985. The evolution of self-fertilization and inbreeding depression in plants. I. Genetic models. *Evolution* **39**, 24–40.

Laverty, T. M. 1992. Plant interactions for pollinator visits: a test of the magnet species effect. *Oecologia* **89**, 502–508.

Lehtonen, J. and Whitehead, M. R. 2014. Sexual deception: coevolution or inescapable exploitation? *Current Zoology* **60**, 52–61.

Li, P., Luo, Y., Bernhardt, P., Kou, Y., and Perner, H. 2008. Pollination of *Cypripedium plectrochilum* (Orchidaceae) by *Lasioglossum* spp. (Halictidae): the roles of generalist attractants versus restrictive floral architecture. *Plant Biology* **10**, 220–230.

Little, R. J. 1983. A review of floral food deception mimicries with comments on floral mutualisms. In: *Handbook of Experimental Pollination Biology* (ed. by Jones, C. E. and Little, R. J.), pp. 294–309. New York: Van Nostrand Reinhold.

Lloyd, D. G. and Barrett, S. C. H. (eds.) 1996. *Floral Biology: Studies on Floral Evolution in Animal-pollinated Plants.* Chapman and Hall, New York.

López- Portillo, J., Eguiarte, L. E., and Montaña, C. 1993. Nectarless honey mesquites. *Functional Ecology* **7**, 452–461.

Lunau, K. 1990. Colour saturation triggers innate reactions to flower signals: flower dummy experiments with bumblebees. *Journal of Comparative Physiology. A, Sensory, Neural, and Behavioral Physiology* **166**, 827–834.

Lunau, K. 1992. Limits to color learning in a flower-visiting hoverfly, *Eristalix tenax* L (Syrphidae, Diptera). *European Journal of Neuroscience Supplement* **5**, 103.

Lunau, K. 2000. The ecology and evolution of visual pollen signals. *Plant Systematics and Evolution* **222**, 89–111.

Lunau, K. and Maier, E. J. 1995. Innate color preferences of flower visitors. *Journal of Comparative Physiology. A, Sensory, Neural, and Behavioral Physiology* **177**, 1–19.

Lunau, K., Wacht, S., and Chittka, L. 1996. Colour choices of naive bumble bees and their implications for colour perception. *Journal of Comparative Physiology. A, Sensory, Neural, and Behavioral Physiology* **178**, 477–489.

Luyt, R. and Johnson, S. D. 2002. Postpollination nectar reabsorption and its implications for fruit quality in an epiphytic orchid. *Biotropica* **34**, 442–446.

Ma, X., et al. 2016. The functional significance of complex floral colour pattern in a food-deceptive orchid. *Functional Ecology* **30**, 721–732.

Maad, J. 2000. Phenotypic selection in hawkmoth-pollinated *Platanthera bifolia*: targets and fitness surfaces. *Evolution* **54**, 112–123.

Maia, R., Eliason, C. M., Bitton, P.-P., Doucet, S. M., and Shawkey, M. D. 2013. pavo: an R package for the analysis, visualization and organization of spectral data. *Methods in Ecology and Evolution* **4**, 906–913.

Majetic, C. J., Raguso, R. A., Tonsor, S. J., and Ashman, T.-L. 2007. Flower color–flower scent associations in polymorphic *Hesperis matronalis* (Brassicaceae). *Phytochemistry* **68**, 865–874.

Makino, T. T. and Sakai, S. 2007. Experience changes pollinator responses to floral display size: from size-based to reward-based foraging. *Functional Ecology* **21**, 854–863.

Mallet, J. 1999. Causes and consequences of a lack of co-evolution in Mullerian mimicry. *Evolutionary Ecology* **13**, 777–806.

Mallet, J. 2015. New genomes clarify mimicry evolution. *Nature Genetics* **47**, 306–307.

Mallet, J. and Joron, M. 1999. Evolution of diversity in warning color and mimicry: polymorphisms, shifting balance, and speciation. *Annual Review of Ecology and Systematics* **30**, 201–233.

Maloof, J. E. 2001. The effects of a bumble bee nectar robber on plant reproductive success and pollinator behavior. *American Journal of Botany* **88**, 1960–1965.

Mant, J., Bower, C. C., Weston, P. H., and Peakall, R. 2005a. Phylogeography of pollinator-specific sexually deceptive *Chiloglottis* taxa (Orchidaceae): evidence for sympatric divergence? *Molecular Ecology* **14**, 3067–3076.

Mant, J., Peakall, R., and Schiestl, F. P. 2005b. Does selection on floral odor promote differentiation among populations and species of the sexually deceptive orchid genus *Ophrys*? *Evolution* **59**, 1449–1463.

Marden, J. H. 1984. Remote perception of floral nectar by bumblebees. *Oecologia* **64**, 232–240.

Marino, P., Raguso, R., and Goffinet, B. 2009. The ecology and evolution of fly dispersed dung mosses (Family Splachnaceae): manipulating insect behaviour through odour and visual cues. *Symbiosis* **47**, 61–76.

Marloth, R. 1907. Notes on the morphology and biology of *Hydnora africana*. *Transactions of the South African Philosophical Society* **16**, 465–468.

Martos, F., Cariou, M.-L., Pailler, T., Fournel, J., Bytebier, B., and Johnson, S. D. 2015. Chemical and morphological filters in a specialized floral mimicry system. *New Phytologist* **207**, 225–234.

Meekers, T. and Honnay, O. 2011. Effects of habitat fragmentation on the reproductive success of the nectar-producing orchid *Gymnadenia conopsea* and the nectar-less *Orchis mascula*. *Plant Ecology* **212**, 1791–1801.

Menz, M. H. M., Phillips, R. D., Anthony, J. M., Bohman, B., Dixon, K. W., and Peakall, R. 2015. Ecological and genetic evidence for cryptic ecotypes in a rare sexually deceptive orchid, *Drakaea elastica*. *Botanical Journal of the Linnean Society* **177**, 124–140.

Menz, M. H. M., Phillips, R. D., Dixon, K. W., Peakall, R., and Didham, R. K. 2013. Mate-searching behaviour of common and rare wasps and the implications for pollen movement of the sexually deceptive orchids they pollinate. *PLoS One* **8**(3), e59111.

Menzel, R. 1985. Learning in honey bees in an ecological and behavioral context. *Fortschritte der Zoologie* **31**, 55–74.

Menzel, R., Greggers, U., and Hammer, M. 1993. Functional organization of appetitive learning and memory in a generalist pollinator, the honey bee. In: *Insect Learning: Ecological and Evolutionary Perspectives* (ed. by Papaj, D. R. and Lewis, A. C.), pp. 79–125. New York: Chapman and Hall.

Meve, U. and Liede, S. 1994. Floral biology and pollination in stapeliads—new results and a literature review. *Plant Systematics and Evolution* **192**, 99–116.

Midgley, J. J., White, J. D. M., Johnson, S. D., and Bronner, G. N. 2015. Faecal mimicry by seeds ensures dispersal by dung beetles. *Nature Plants* **1**, art no. 15141.

Milet-Pinheiro, P., Ayasse, M., and Dötterl, S. 2015. Visual and olfactory floral cues of *Campanula* (Campanulaceae) and their significance for host recognition by an oligolectic bee pollinator. *PLoS One* **10**(6), e0128577.

Milet-Pinheiro, P., Ayasse, M., Dobson, H. E. M., Schlindwein, C., Francke, W., and Dötterl, S. 2013. The chemical basis of host-plant recognition in a specialized bee pollinator. *Journal of Chemical Ecology* **39**, 1347–1360.

Milet-Pinheiro, P., Ayasse, M., Schlindwein, C., Dobson, H. E. M., and Dötterl, S. 2012. Host location by visual and olfactory floral cues in an oligolectic bee: innate and learned behavior. *Behavioral Ecology* **23**, 531–538.

Moggridge, J. T. 1869. *Über* Ophrys insectifera *L. (part.)*. Dresden: Blochmann.

Molvray, M., Kores, P., and Chase, M. W. 2000. Polyphyly of mycoheterotrophic orchids and functional influences on floral and molecular characters. In: *Monocots: Systematics and Evolution* (ed. by Wilson, K. L. and Morrison, D. A.), pp. 441–448. Melbourne: CSIRO.

Mondragón-Palomino, M. and Theißen, G. 2008. MADS about the evolution of orchid flowers. *Trends in Plant Science* **13**, 51–59.

Mondragon-Palomino, M. and Theißen, G. 2011. Conserved differential expression of paralogous DEFICIENS- and GLOBOSA-like MADS-box genes in the flowers of Orchidaceae: refining the "orchid code." *Plant Journal* **66**, 1008–1019.

Mondragon-Palomino, M., Hiese, L., Haerter, A., Koch, M. A., and Theißen, G. 2009. Positive selection and ancient duplications in the evolution of class B floral homeotic genes of orchids and grasses. *BMC Evolutionary Biology* **9**, 81.

Montalvo, A. M. and Ackerman, J. D. 1987. Limitations to fruit production in *Ionopsis utricularoides* (Orchidaceae). *Biotropica* **19**, 24–31.

Moorhouse, J. E., Yeadon, R., Beevor, P. S., and Nesbit, F. B. 1969. Method for use in studies of insect chemical communication. *Nature* **223**, 1174–1175.

More, M., Cocucci, A. A., and Raguso, R. A. 2013. The importance of oligosulfides in the attraction of fly pollinators to the brood-site deceptive species *Jaborosa rotacea* (solanaceae). *International Journal of Plant Sciences* **174**, 863–876.

Naef, A., Roy, B. A., Kaiser, R., and Honegger, R. 2002. Insect-mediated reproduction of systemic infections by *Puccinia arrhenatheri* on *Berberis vulgaris*. *New Phytologist* **154**, 717–730.

Ne'eman, G., Shavit, O., Shaltiel, L., and Shmida, A. 2006. Foraging by male and female solitary bees with implications for pollination. *Journal of Insect Behavior* **19**, 383–401.

Neiland, M. R. M. and Wilcock, C. C. 1998. Fruit set, nectar reward, and rarity in the Orchidaceae. *American Journal of Botany* **85**, 1657–1671.

Neiland, M. R. M. and Wilcock, C. C. 1999. The presence of heterospecific pollen on stigmas of nectariferous and nectarless orchids and its consequences for their reproductive success. *Protoplasma* **208**, 65–75.

Neiland, M. R. M. and Wilcock, C. C. 2000. Effects of pollinator behaviour on pollination of nectarless orchids: flo-

ral mimicry and interspecific hybridization. In: *Monocots: Systematics and Evolution* (ed. by Wilson, K. L. and Morrison, D. A.), pp. 318–325. Melbourne: CSIRO.

Nepi, M. and Stpiczynska, M. 2007. Nectar resorption and translocation in *Cucurbita pepo* L. and *Platanthera chlorantha* Custer (Rchb.). *Plant Biology* **9**, 93–100.

Newman, E., Anderson, B., and Johnson, S. D. 2012. Flower colour adaptation in a mimetic orchid. *Proceedings of the Royal Society B: Biological Sciences* **279**, 2309–2313.

Ngugi, H. K. and Scherm, H. 2006. Mimicry in plant-parasitic fungi. *FEMS Microbiology Letters* **257**, 171–176.

Nierenberg, L. 1972. The mechanism for the maintenance of species integrity in sympatrically occurring equitant *Oncidium* in the Caribbean. *American Orchid Society Bulletin* **41**, 873–882.

Nilsson, L. A. 1980. The pollination ecology of *Dactylorhiza sambucina* (Orchidaceae). *Botaniska Notiser* **133**, 368–385.

Nilsson, L. A. 1983a. Anthecology of *Orchis mascula* (Orchidaceae). *Nordic Journal of Botany* **3**, 157–179.

Nilsson, L. A. 1983b. Mimesis of bellflower (*Campanula*) by the red helleborine orchid *Cephalanthera rubra*. *Nature* **305**, 799–800.

Nilsson, L. A. 1992. Orchid pollination biology. *Trends in Ecology and Evolution* **7**, 255–259.

Nilsson, L. A., Rabakonandrianina, E., and Pettersson, B. 1992. Exact tracking of pollen transfer and mating in plants. *Nature* **360**, 666–668.

Nishikawa, H., et al. 2015. A genetic mechanism for female-limited Batesian mimicry in *Papilio* butterfly. *Nature Genetics* **47**, 405–409.

Oaten, A., Pearce, C. E. M., and Smyth, M. E. B. 1975. Batesian mimicry and signal detection theory. *Bulletin of Mathematical Biology* **37**, 367–387.

O'Connell, L. M. and Johnston, M. O. 1998. Male and female pollination success in a deceptive orchid, a selection study. *Ecology* **79**, 1246–1260.

Oelschlägel, B., et al. 2015. The betrayed thief—the extraordinary strategy of *Aristolochia rotunda* to deceive its pollinators. *New Phytologist* **206**, 342–351.

O'Hanlon, J. C. 2014. The roles of colour and shape in pollinator deception in the orchid mantis *Hymenopus coronatus*. *Ethology* **120**, 652–661.

O'Hanlon, J. C., Herberstein, M. E., and Holwell, G. I. 2015. Habitat selection in a deceptive predator: maximizing resource availability and signal efficacy. *Behavioral Ecology* **26**, 194–199.

Ollerton, J. and Raguso, R. A. 2006. The sweet stench of decay. *New Phytologist* **172**, 382–385.

Ollerton, J., Masinde, S., Meve, U., Picker, M., and Whittington, A. 2009. Fly pollination in *Ceropegia* (Apocynaceae: Asclepiadoideae): biogeographic and phylogenetic perspectives. *Annals of Botany* **103**, 1501–1514.

Onda, Y., et al. 2008. Functional coexpression of the mitochondrial alternative oxidase and uncoupling protein underlies thermoregulation in the thermogenic florets of skunk cabbage. *Plant Physiology* **146**, 636–645.

Onda, Y., et al. 2015. Transcriptome analysis of thermogenic *Arum concinnatum* reveals the molecular components of floral scent production. *Scientific Reports* **5**, 8753.

Osche, G. 1979. Zur Evolution optischer Signale bei Blütenpflanzen. *Biologie in unserer Zeit* **9**, 161–170.

Paczkowski, S., Nicke, S., Ziegenhagen, H., and Schuetz, S. 2015. Volatile emission of decomposing pig carcasses (*Sus scrofa domesticus* L.) as an indicator for the postmortem interval. *Journal of Forensic Sciences* **60**, S130–S137.

Paczkowski, S. and Schuetz, S. 2011. Post-mortem volatiles of vertebrate tissue. *Applied Microbiology and Biotechnology* **91**, 917–935.

Pan, I. C., et al. 2012. Complete chloroplast genome sequence of an orchid model plant candidate: *Erycina pusilla* apply in tropical *Oncidium* breeding. *PLoS One* **7**(4), e34738.

Pansarin, E. R. 2008. Reproductive biology and pollination of *Govenia utriculata*: a syrphid fly orchid pollinated through a pollen-deceptive mechanism. *Plant Species Biology* **23**, 90–96.

Pansarin, E. R., Salatino, A., Pansarin, L. M., and Sazima, M. 2012. Pollination systems in Pogonieae (Orchidaceae: Vanilloideae): a hypothesis of evolution among reward and rewardless flowers. *Flora* **207**, 849–861.

Papadopulos, A. S. T., et al. 2013. Convergent evolution of floral signals underlies the success of Neotropical orchids. *Proceedings of the Royal Society B: Biological Sciences* **280**, 20130960.

Pasteur, G. 1982. A classificatory review of mimicry systems. *Annual Review of Ecology and Systematics* **13**, 169–199.

Paulus, H. F. and Gack, C. 1990. Pollinators as prepollinating isolation factors: evolution and speciation in *Ophrys* (Orchidaceae). *Israel Journal of Botany* **39**, 43–97.

Peakall, R. 1990. Responses of male *Zaspilothynnus trilobatus* Turner wasps to females and the sexually deceptive orchid it pollinates. *Functional Ecology* **4**, 159–167.

Peakall, R. and Beattie, A. J. 1996. Ecological and genetic consequences of pollination by sexual deception in the orchid *Caladenia tentactulata*. *Evolution* **50**, 2207–2220.

Peakall, R. and Handel, S. N. 1993. Pollinators discriminate among floral heights of a sexually deceptive orchid: implications for selection. *Evolution* **47**, 1681–1687.

Peakall, R. and Schiestl, F. P. 2004. A mark–recapture study of male *Colletes cunicularius* bees: implications for pollination by sexual deception. *Behavioral Ecology and Sociobiology* **56**, 579–584.

Peakall, R. and Whitehead, M. R. 2014. Floral odour chemistry defines species boundaries and underpins strong reproductive isolation in sexually deceptive orchids. *Annals of Botany* **113**, 341–355.

Peakall, R., Angus, C. J., and Beattie, A. J. 1990. The significance of ant and plant traits for ant pollination in *Leporella fimbriata*. *Oecologia* **84**, 457–460.

Peakall, R., Bower, C. C., Logan, A. E., and Nicol, H. I. 1997. Confirmation of the hybrid origin of *Chiloglottis* × *Pescottiana* (Orchidaceae: Diurideae). 1. Genetic and morphometric evidence. *Australian Journal of Botany* **45**, 839–855.

Peakall, R., et al. 2010. Pollinator specificity, floral odour chemistry and the phylogeny of Australian sexually deceptive *Chiloglottis* orchids: implications for pollinator-driven speciation. *New Phytologist* **188**, 437–450.

Pellegrino, G., Bellusci, F., and Musacchio, A. 2005a. Evidence of post-pollination barriers among three colour morphs of the deceptive orchid *Dactylorhiza sambucina* (L.) Soo. *Sexual Plant Reproduction* **18**, 179–185.

Pellegrino, G., Bellusci, F., and Musacchio, A. 2008. Double floral mimicry and the magnet species effect in dimorphic co-flowering species, the deceptive orchid *Dactylorhiza sambucina* and rewarding *Viola aethnensis*. *Preslia* **80**, 411–422.

Pellegrino, G., Caimi, D., Noce, M. E., and Musacchio, A. 2005b. Effects of local density and flower colour polymorphism on pollination and reproduction in the rewardless orchid *Dactylorhiza sambucina* (L.) Soò. *Plant Systematics and Evolution* **251**, 119–129.

Pellissier, L., Vittoz, P., Internicola, A. I., and Gigord, L. D. B. 2010. Generalized food-deceptive orchid species flower earlier and occur at lower altitudes than rewarding ones. *Journal of Plant Ecology* **3**, 243–250.

Pellmyr, O. 1986. The pollination ecology of two nectarless *Cimicifuga* sp. (Ranunculaceae) in North America. *Nordic Journal of Botany* **6**, 713–723.

Pemberton, R. 2011. Pollination studies in phragmipediums: flower fly (Syrphidae) pollination and mechanical self pollination (autogamy) in *Phragmipedium* species (Cypripedioideae). *Orchids* **80**, 364–367.

Penney, H. D., Hassall, C., Skevington, J. H., Abbott, K. R., and Sherratt, T. N. 2012. A comparative analysis of the evolution of imperfect mimicry. *Nature* **483**, 461–464.

Peter, C. I. and Johnson, S. D. 2006. Doing the twist: a test of Darwin's cross-pollination hypothesis for pollination reconfiguration. *Biology Letters* **2**, 65–68.

Peter, C. I. and Johnson, S. D. 2008. Mimics and magnets: the importance of color and ecological facilitation in floral deception. *Ecology* **89**, 1583–1595.

Peter, C. I. and Johnson, S. D. 2009. Reproductive biology of *Acrolophia cochlearis* (Orchidaceae): estimating rates of cross-pollination in epidendroid orchids. *Annals of Botany* **104**, 573–581.

Peter, C. I. and Johnson, S. D. 2013. Generalized food deception: colour signals and efficient pollen transfer in bee-pollinated species of *Eulophia* (Orchidaceae). *Botanical Journal of the Linnean Society* **171**, 713–729.

Peter, C. I. and Johnson, S. D. 2014. A pollinator shift explains floral divergence in an orchid species complex in South Africa. *Annals of Botany* **113**, 277–288.

Petterson, B. and Nilsson, L. A. 1993. Floral variation and deceit pollination in *Polystachya rosea* (Orchidaceae) on an inselberg in Madagascar. *Opera Botanica* **121**, 237–245.

Pfennig, D. W., Akcali, C. K., and Kikuchi, D. W. 2015. Batesian mimicry promotes pre- and postmating isolation in a snake mimicry complex. *Evolution* **69**, 1085–1090.

Pfunder, M. and Roy, B. A. 2000. Pollinator-mediated interactions between a pathogenic fungus, *Uromyces pisi* (Pucciniaceae), and its host plant, *Euphorbia cyparissias* (Euphorbiaceae). *American Journal of Botany* **87**, 48–55.

Phillips, R. D., Dixon, K. W., and Peakall, R. 2012. Low population genetic differentiation in the Orchidaceae: implications for the diversification of the family. *Molecular Ecology* **21**, 5208–5220.

Phillips, R. D., et al. 2014a. Specialized ecological interactions and plant species rarity: the role of pollinators and mycorrhizal fungi across multiple spatial scales. *Biological Conservation* **169**, 285–295.

Phillips, R. D., et al. 2014b. Caught in the act: pollination of sexually deceptive trap-flowers by fungus gnats in *Pterostylis* (Orchidaceae). *Annals of Botany* **113**, 629–641.

Phillips, R. D., Xu, T., Hutchinson, M. F., Dixon, K. W., and Peakall, R. 2013. Convergent specialization—the sharing of pollinators by sympatric genera of sexually deceptive orchids. *Journal of Ecology* **101**, 826–835.

Pleasants, J. M. and Chaplin, S. J. 1983. Nectar production rates of *Asclepias quadrifolia*: causes and consequences of individual variation. *Oecologia* **59**, 232–238.

Plepys, D., Ibarra, F., and Löfstedt, C. 2002. Volatiles from flowers of *Platanthera bifolia* (Orchidaceae) attractive to the silver Y moth, *Autographa gamma* (Lepidoptera: Noctuidae). *Oikos* **99**, 69–74.

Pohl, M., Watolla, T., and Lunau, K. 2008. Anther-mimicking floral guides exploit a conflict between innate preference and learning in bumblebees (*Bombus terrestris*). *Behavioral Ecology and Sociobiology* **63**, 295–302.

Policha, T. 2014. Pollination biology of the mushroom-mimicking orchid genus *Dracula*. PhD Dissertation. Eugene, OR: University of Oregon.

Policha, T., Davis, A. R., Barnadas, M., Dentinger, B. M., Raguso, R. A., and Roy, B. 2016. Disentangling visual and olfactory signals in mushroom-mimicking *Dracula* orchids using realistic 3D printed flowers. *New Phytologist* **210**, 1058–1071.

Poulton, E. B. 1890. *The Colours of Animals: Their Meaning and Use Especially Considered in the Case of Insects*. London: Kegan Paul, Trench, Trubner & Co. Ltd.

Pouyanne, A. 1917. Le fecondation des *Ophrys* par les insectes. *Bulletin de la Société d'Histoire Naturelle d'Afrique du Nord* **8**, 6–7.

Powell, E. A. and Jones, C. E. 1983. Floral mutualism in *Lupinus benthamii* (Fabaceae) and *Delphinium parryi* (Ranunculaceae). In: *Handbook of Experimental Pollination Biology* (ed. by Jones, C. E. and Little, R. J.), pp. 310–329. New York: Van Nostrand Reinhold.pp. 9–28. Amsterdam: Elsevier.

Proches, S. and Johnson, S. D. 2009. Beetle pollination of the fruit-scented cones of the South African cycad *Stangeria eriopus*. *American Journal of Botany* **96**, 1722–1730.

Punekar, S. A. and Kumaran, K. P. N. 2010. Pollen morphology and pollination ecology of *Amorphophallus* species from North Western Ghats and Konkan region of India. *Flora* **205**, 326–336.

Punnet, R. C. 1915. *Mimicry in Butterflies*: Cambridge: Cambridge University Press.

Pyke, G. H. 1991. What does it cost a plant to produce nectar? *Nature* **350**, 58–59.

Raguso, R. A. 2001. Floral scent, olfaction, and scent-driven foraging behavior. In: *Cognitive Ecology of Pollination* (ed. by Chittka, L. and Thomson, J. D.), pp. 83–105. Cambridge: Cambridge University Press.

Raguso, R. A. 2004. Why are some floral nectars scented? *Ecology* **85**, 1486–1494.

Raguso, R. A. 2008. Start making scents: the challenge of integrating chemistry into pollination ecology. *Entomologia Experimentalis et Applicata* **128**, 196–207.

Raguso, R. A. and Roy, B. A. 1998. "Floral" scent production by *Puccinia* rust fungi that mimic flowers. *Molecular Ecology* **7**, 1127–1136.

Raine, N. E. and Chittka, L. 2007. The adaptive significance of sensory bias in a foraging context: floral colour preferences in the bumblebee *Bombus terrestris*. *PLoS One* **2**(6), e556.

Rakosy, D., Streinzer, M., Paulus, H. F., and Spaethe, J. 2012. Floral visual signal increases reproductive success in a sexually deceptive orchid. *Arthropod–Plant Interactions* **6**, 671–681.

Ramirez, S. R., et al. 2011. Asynchronous diversification in a specialized plant–pollinator mutualism. *Science* **333**, 1742–1746.

Rathcke, B. 1983. Competition and facilitation among plants for pollination. In: *Pollination Biology* (Ed. by Real, L.), pp. 305–329. Orlando, Florida: Academic Press.

Rausher, M. D. 2008. Evolutionary transitions in floral color. *International Journal of Plant Sciences* **169**, 7–21.

Real, L. A. and Rathcke, B. J. 1991. Individual variation in nectar production and its effect on fitness in *Kalmia latifolia*. *Ecology* **72**, 149–155.

Renner, S. S. 2006. Rewardless flowers in the angiosperms and the role of insect cognition in their evolution. In: *Plant–pollinator Interactions: From Specialization to Generalization* (ed. by Waser, N. M. and Ollerton, J.), pp. 123–144. Chicago: Chicago University Press.

Renner, S. S. 2006. Rewardless flowers in the angiosperms and the role of insect cognition in their evolution. In: *Plant–Pollinator Interactions* (ed. by Waser, N. M. and Ollerton, J.), pp. 123–144. Chicago: University of Chicago Press.

Renner, S. S. and Feil, J. P. 1993. Pollinators of tropical dioecious angiosperms. *American Journal of Botany* **80**, 1100–1107.

Renoult, J. P., Kelber, A., and Schaefer, M. H. 2015. Colour spaces in ecology and evolutionary biology. *Biological Reviews* doi: 10.1111/brv.12230.

Ries, L. and Mullen, S. P. 2008. A rare model limits the distribution of its more common mimic: a twist on frequency-dependent Batesian mimicry. *Evolution* **62**, 1798–1803.

Riffell, J. A., Lei, H., Abrell, L., and Hildebrand, J. G. 2013. Neural basis of a pollinator's buffet: olfactory specialization and learning in *Manduca sexta*. *Science* **339**, 200–204.

Rodriguez-Girones, M. A. and Santamaria, L. 2004. Why are so many bird flowers red? *PLoS Biology* **2**, 1515–1519.

Romero, G. A. and Nelson, C. E. 1986. Sexual dimorphism in *Catasetum* orchids: forcible pollen emplacement and male competition. *Science* **232**, 1538–1540.

Rothschild, M. 1981. The mimicrats must move with the times. *Biological Journal of the Linnean Society* **16**, 21–23.

Roy, B. A. 1993. Floral mimicry by a plant pathogen. *Nature* **362**, 56–58.

Roy, B. A. 1996. A plant pathogen influences pollinator behavior and may influence reproduction of nonhosts. *Ecology* **77**, 2445–2457.

Roy, B. A. 2001. Patterns of association between crucifers and their flower-mimic pathogens: host jumps are more common than coevolution or cospeciation. *Evolution* **55**, 41–53.

Roy, B. A. and Raguso, R. A. 1997. Olfactory versus visual cues in a floral mimicry system. *Oecologia* **109**, 414–426.

Roy, B. A. and Widmer, A. 1999. Floral mimicry: a fascinating yet poorly understood phenomenon. *Trends in Plant Science* **4**, 325–330.

Ruxton, G. D. and Schaefer, H. M. 2011. Alternative explanations for apparent mimicry. *Journal of Ecology* **99**, 899–904.

Ruxton, G. D., Sherratt, T. N., and Speed, M. P. 2004. *Avoiding Attack: The Evolutionary Ecology of Crypsis, Warning Signals and Mimicry*. Oxford: Oxford University Press.

Sachse, S., Rappert, A., and Galizia, C. G. 1999. The spatial representation of chemical structures in the antennal lobe of honeybees: steps towards the olfactory code. *European Journal of Neuroscience* **11**, 3970–3982.

Sakai, S. 2002. *Aristolochia* spp. (Aristolochiaceae) pollinated by flies breeding on decomposing flowers in Panama. *American Journal of Botany* **89**, 527–534.

Sakai, S. and Inoue, T. 1999. A new pollination system: dung-beetle pollination discovered in *Orchidantha inouei* (Lowiaceae, Zingiberales) in Sarawak, Malaysia. *American Journal of Botany* **86**, 56–61.

Salzmann, C. C., Cozzolino, S., and Schiestl, F. P. 2007a. Floral scent in food-deceptive orchids: species specificity and sources of variability. *Plant Biology* **9**, 720–729.

Salzmann, C. C., Nardella, A. M., Cozzolino, S., and Schiestl, F. P. 2007b. Variability in floral scent in rewarding and deceptive orchids: the signature of pollinator-imposed selection? *Annals of Botany* **100**, 757–765.

Salzmann, C. C. and Schiestl, F. P. 2007. Odour and colour polymorphism in the food-deceptive orchid *Dactylorhiza romana*. *Plant Systematics and Evolution* **267**, 37–45.

Sazima, M. and Sazima, I. 1989. Oil-gathering bees visit flowers of eglandular morphs of the oil-producing Malphigiaceae. *Botanica Acta* **102**, 106–111.

Schaefer, H. M. and Ruxton, G. D. 2009. Deception in plants: mimicry or perceptual exploitation? *Trends in Ecology and Evolution* **24**, 676–685.

Schäffler, I., et al. 2015. Diacetin, a reliable cue and private communication channel in a specialized pollination system. *Scientific Reports* **5**, art. no. 12779.

Schemske, D. W. 1981. Floral convergence and pollinator sharing in two bee-pollinated tropical herbs. *Ecology* **62**, 946–954.

Schemske, D. W. and Ågren, J. 1995. Deceit pollination and selection on female flower size in *Begonia involucrata*: an experimental approach. *Evolution* **49**, 207–214.

Schemske, D. W., Ågren, J., and Le Corff, J. 1996. Deceit pollination in the monoecious, Neotropical herb *Begonia involucrata*: an experimental approach. In: *Floral Biology: Studies on Floral Evolution in Animal-pollinated Plants* (ed. by Lloyd, D. G. and Barrett, S. C. H.), pp. 292–318. New York: Chapman and Hall.

Schiestl, F. P. 2004. Floral evolution and pollinator mate choice in a sexually deceptive orchid. *Journal of Evolutionary Biology* **17**, 67–75.

Schiestl, F. P. 2005. On the success of a swindle: pollination by deception in orchids. *Naturwissenschaften* **92**, 255–264.

Schiestl, F. P. and Ayasse, M. 2001. Post-pollination emission of a repellent compound in a sexually deceptive orchid: a new mechanism for maximising reproductive success? *Oecologia* **126**, 531–534.

Schiestl, F. P. and Ayasse, M. 2002. Do changes in floral odor cause speciation in sexually deceptive orchids? *Plant Systematics and Evolution* **234**, 111–119.

Schiestl, F. P. and Cozzolino, S. 2008. Evolution of sexual mimicry in the orchid subtribe orchidinae: the role of preadaptations in the attraction of male bees as pollinators. *BMC Evolutionary Biology* **8**, 27.

Schiestl, F. P. and Dötterl, S. 2012. The evolution of floral scent and olfactory preferences in pollinators: coevolution or pre-existing bias? *Evolution* **66**, 2042–2055.

Schiestl, F. P. and Marion-Poll, F. 2002. Detection of physiologically active flower volatiles using gas chromatography coupled with electroantennography. In: *Analysis of Taste and Aroma* (ed. by Jackson, J. F. and Linskens, H. F.), pp. 173–198. Berlin: Springer.

Schiestl, F. P. and Peakall, R. 2005. Two orchids attract different pollinators with the same floral odour compound: ecological and evolutionary implications. *Functional Ecology* **19**, 674–680.

Schiestl, F. P. and Schlüter, P. M. 2009. Floral isolation, specialized pollination, and pollinator behavior in orchids. *Annual Review of Entomology* **54**, 425–446.

Schiestl, F. P., et al. 1999. Orchid pollination by sexual swindle. *Nature* **399**, 421–422.

Schiestl, F. P., et al. 2000. Sex pheromone mimicry in the early spider orchid (*Ophrys sphegodes*): patterns of hydrocarbons as the key mechanism for pollination by sexual deception. *Journal of Comparative Physiology. A, Sensory, Neural, and Behavioral Physiology* **186**, 567–574.

Schiestl, F. P., et al. 2003. The chemistry of sexual deception in an orchid–wasp pollination system. *Science* **302**, 437–438.

Schiestl, F. P., et al. 2006. Evolution of "pollinator"-attracting signals in fungi. *Biology Letters* **2**, 401–404.

Schiml, S. and Puchta, H. 2016. Revolutionizing plant biology: multiple ways of genome engineering by CRISPR/Cas. *Plant Methods* **12**, 8.

Schlüter, P. M. and Schiestl, F. P. 2008. Molecular mechanisms of floral mimicry in orchids. *Trends in Plant Science* **13**, 228–235.

Schlüter, P. M., et al. 2011. Stearoyl-acyl carrier protein desaturases are associated with floral isolation in sexually deceptive orchids. *Proceedings of the National Academy of Sciences of the United States of America* **108**, 5696–5701.

Schmitt, U. and Bertsch, A. 1990. Do foraging bumblebees scent-mark food sources and does it matter? *Oecologia* **82**, 137–144.

Schulz, S. and Dickschat, J. S. 2007. Bacterial volatiles: the smell of small organisms. *Natural Product Reports* **24**, 814–842.

Scopece, G., Cozzolino, S., Johnson, S. D., and Schiestl, F. P. 2010. Pollination efficiency and the evolution of specialized deceptive pollination systems. *The American Naturalist* **175**, 98–105.

Scopece, G., Juillet, N., Muller, A., Schiestl, F. P., and Cozzolino, S. 2009. Pollinator attraction in *Anacamptis papilionacea* (Orchidaceae): a food or a sex promise? *Plant Species Biology* **24**, 109–114.

Scopece, G., Musacchio, A., Widmer, A., and Cozzolino, S. 2007. Patterns of reproductive isolation in Mediterranean deceptive orchids. *Evolution* **61**, 2623–2642.

Sedeek, K. E. M., et al. 2014. Genic rather than genome-wide differences between sexually deceptive *Ophrys* orchids with different pollinators. *Molecular Ecology* **23**, 6192–6205.

Seymour, R. S. 2010. Scaling of heat production by thermogenic flowers: limits to floral size and maximum rate of respiration. *Plant Cell and Environment* **33**, 1474–1485.

Seymour, R. S., Gibernau, M., and Ito, K. 2003. Thermogenesis and respiration of inflorescences of the dead horse arum *Helicodiceros muscivorus*, a pseudo-thermoregulatory aroid associated with fly pollination. *Functional Ecology* **17**, 886–894.

Seymour, R. S., Ito, K., Umekawa, Y., Matthews, P. D. G., and Pirintsos, S. A. 2015. The oxygen supply to thermogenic flowers. *Plant Cell and Environment* **38**, 827–837.

Seymour, R. S., Ito, Y., Onda, Y., and Ito, K. 2009. Effects of floral thermogenesis on pollen function in Asian skunk cabbage *Symplocarpus renifolius*. *Biology Letters* **5**, 568–570.

Seymour, R. S. and Schultze-Motel, P. 1997. Heat-producing flowers. *Endeavour* **21**, 125–129.

Shanklin, J. and Somerville, C. 1991. Stearoyl-acyl-carrier-protein desaturase from higher-plants is structurally unrelated to the animal and fungal homologs. *Proceedings of the National Academy of Sciences of the United States of America* **88**, 2510–2514.

Sherratt, T. N. 2002. The evolution of imperfect mimicry. *Behavioral Ecology* **13**, 821–826.

Sherratt, T. N. 2008. The evolution of Mullerian mimicry. *Naturwissenschaften* **95**, 681–695.

Shi, J., Cheng, J., Luo, D., Shangguan, F. Z., and Luo, Y. B. 2007. Pollination syndromes predict brood-site deceptive pollination by female hoverflies in *Paphiopedilum dianthum* (Orchidaceae). *Acta Phytotaxonomica Sinica* **45**, 551–560.

Shi, J., Luo, Y. B., Bernhardt, P., Ran, J. C., Liu, Z. J., and Zhou, Q. 2009. Pollination by deceit in *Paphiopedilum barbigerum* (Orchidaceae): a staminode exploits the innate colour preferences of hoverflies (Syrphidae). *Plant Biology* **11**, 17–28.

Shuttleworth, A. and Johnson, S. D. 2010. The missing stink: sulphur compounds can mediate a shift between fly and wasp pollination systems. *Proceedings of the Royal Society B: Biological Sciences* **277**, 2811–2819.

Shykoff, J. A. and Bucheli, E. 1995. Pollinator visitation patterns, floral rewards and the probability of transmission of *Microbotryum violaceae*, a venereal disease of plants. *Journal of Ecology* **83**, 189–198.

Simonds, V. and Plowright, C. M. S. 2004. How do bumblebees first find flowers? Unlearned approach responses and habituation. *Animal Behaviour* **67**, 379–386.

Simpson, B. B. and Neff, J. L. 1983. Evolution and diversity of floral rewards. In: *Handbook of Experimental Pollination Biology* (ed. by Jones, C. E. and Little, R. J.), pp. 142–159. New York: Van Nostrand Reinhold.

Sinn, B. T., Kelly, L. M., and Freudenstein, J. V. 2015. Putative floral brood-site mimicry, loss of autonomous selfing, and reduced vegetative growth are significantly correlated with increased diversification in *Asarum* (Aristolochiaceae). *Molecular Phylogenetics and Evolution* **89**, 194–204.

Skelhorn, J., Halpin, C. G., and Rowe, C. 2016. Learning about aposematic prey. *Behavioral Ecology* doi: 10.1093/beheco/arw009 [online early view].

Skelhorn, J., Rowland, H. M., and Ruxton, G. D. 2010. The evolution and ecology of masquerade. *Biological Journal of the Linnean Society* **99**, 1–8.

Sleeman, D. P., Jones, P., and Cronin, J. N. 1997. Investigations of an association between the stinkhorn fungus and badger setts. *Journal of Natural History* **31**, 983–992.

Sletvold, N., Grindeland, J. M., and Ågren, J. 2010. Pollinator-mediated selection on floral display, spur length and flowering phenology in the deceptive orchid *Dactylorhiza lapponica*. *New Phytologist* **188**, 385–392.

Sletvold, N., Grindeland, J. M., and Ågren, J. 2013. Vegetation context influences the strength and targets of pollinator-mediated selection in a deceptive orchid. *Ecology* **94**, 1236–1242.

Smith, B. H. 1991. The olfactory memory of the honey bee *Apis mellifera*. 1. Odorant modulation of short-term and intermediate memory after single-trail conditioning. *Journal of Experimental Biology* **161**, 367–382.

Smithson, A. 2001. Pollinator preference, frequency dependence, and floral evolution. In: *Cognitive Ecology of Pollination: Animal Behaviour and Floral Evolution* (ed. by Chittka, L. and Thomson, J. D.), pp. 237–258. Cambridge: Cambridge University Press.

Smithson, A. 2002. The consequences of rewardlessness in orchids: reward-supplementation experiments with *Anacamptis morio* (Orchidaceae). *American Journal of Botany* **89**, 1579–1587.

Smithson, A. 2006. Pollinator limitation and inbreeding depression in orchid species with and without nectar rewards. *New Phytologist* **169**, 419–430.

Smithson, A. 2009. A plant's view of cheating in plant–pollinator mutualisms. *Israel Journal of Plant Sciences* **57**, 151–163.

Smithson, A. and Gigord, L. D. B. 2001. Are there fitness advantages in being a rewardless orchid? Reward supplementation experiments with *Barlia robertiana*. *Proceedings of the Royal Society B: Biological Sciences* **268**, 1435–1441.

Smithson, A. and Gigord, L. D. B. 2003. The evolution of empty flowers revisited. *The American Naturalist* **161**, 537–552.

Smithson, A. and Macnair, M. R. 1996. Frequency-dependent selection by pollinators: mechanisms and consequences with regard to behaviour of bumblebees *Bombus terrestris* (L) (Hymenoptera: Apidae). *Journal of Evolutionary Biology* **9**, 571–588.

Smithson, A. and Macnair, M. R. 1997. Negative frequency-dependent selection by pollinators on artificial flowers without rewards. *Evolution* **51**, 715–723.

Smithson, A., Juillet, N., Macnair, M. R., and Gigord, L. D. B. 2007. Do rewardless orchids show a positive relationship between phenotypic diversity and reproductive success? *Ecology* **88**, 434–442.

Smouse, P. E., Whitehead, M. R., and Peakall, R. 2015. An informational diversity framework, illustrated with sexually deceptive orchids in early stages of speciation. *Molecular Ecology Resources* **15**, 1375–1384.

Soler, C. C. L., Proffit, M., Bessiere, J.-M., Hossaert-McKey, M., and Schatz, B. 2012. Evidence for intersexual chemical mimicry in a dioecious plant. *Ecology Letters* **15**, 978–985.

Soliva, M. and Widmer, A. 2003. Gene flow across species boundaries in sympatric, sexually deceptive *Ophrys* (Orchidaceae) species. *Evolution* **57**, 2252–2261.

Soliva, M., Kocyan, A., and Widmer, A. 2001. Molecular phylogenetics of the sexually deceptive orchid genus *Ophrys* (Orchidaceae) based on nuclear and chloroplast DNA sequences. *Molecular Phylogenetics and Evolution* **20**, 78–88.

Southwick, E. E. 1984. Photosynthate allocation to floral nectar: a neglected energy investment. *Ecology* **65**, 1775–1779.

Spaethe, J., Moser, W. H., and Paulus, H. F. 2007. Increase of pollinator attraction by means of a visual signal in the sexually deceptive orchid, *Ophrys heldreichii* (Orchidaceae). *Plant Systematics and Evolution* **264**, 31–40.

Spaethe, J., Tautz, J., and Chittka, L. 2001. Visual constraints in foraging bumblebees: flower size and color affect search time and flight behavior. *Proceedings of the National Academy of Sciences of the United States of America* **98**, 3898–3903.

Speed, M. P. 1993. Mullerian mimicry and the psychology of predation. *Animal Behaviour* **57**, 203–213.

Speed, M. P. and Turner, J. R. G. 1999. Learning and memory in mimicry: II. Do we understand the mimicry spectrum? *Biological Journal of the Linnean Society* **67**, 281–312.

Sprengel, C. K. 1793. *Das entdeckte Geheimnis der Natur im Bau und in der Befruchtung der Blumen.* Berlin: Friedrich Vieweg dem Aeltern.

Starrett, A. 1993. Adaptive resemblance: a unifying concept for mimicry and crypsis. *Biological Journal of the Linnean Society* **48**, 299–317.

Stebbins, G. L. 1970. Adaptive radiation of reproductive characteristics in angiosperms. I. Pollination mechanisms. *Annual Review of Ecology and Systematics* **1**, 307–326.

Steinebrunner, F., Twele, R., Francke, W., Leuchtmann, A., and Schiestl, F. P. 2008. Role of odour compounds in the attraction of gamete vectors in endophytic *Epichloe* fungi. *New Phytologist* **178**, 401–411.

Steiner, K. E., Whitehead, V. B., and Johnson, S. D. 1994. Floral and pollinator divergence in two sexually deceptive South-African orchids. *American Journal of Botany* **81**, 185–194.

Stensmyr, M. C., Urru, I., Collu, I., Celander, M., Hansson, B. S., and Angioy, A. M. 2002. Rotting smell of dead-horse arum florets. *Nature* **420**, 625–626.

Stökl, J., Brodmann, J., Dafni, A., Ayasse, M., and Hansson, B. S. 2011. Smells like aphids: orchid flowers mimic aphid alarm pheromones to attract hoverflies for pollination. *Proceedings of the Royal Society B: Biological Sciences* **278**, 1216–1222.

Stökl, J., et al. 2010. A deceptive pollination system targeting drosophilids through olfactory mimicry of yeast. *Current Biology* **20**, 1846–1852.

Stökl, J., Schlueter, P. M., Stuessy, T. F., Paulus, H. F., Assum, G., and Ayasse, M. 2008. Scent variation and hybridization cause the displacement of a sexually deceptive orchid species. *American Journal of Botany* **95**, 472–481.

Stoutamire, W. P. 1975. Pseudocopulation in Australian terrestrial orchids. *American Orchid Society Bulletin* **44**, 226–233.

Stoutamire, W. P. 1983. Wasp pollination species of *Caladenia* (Orchidacea) in south-western Australia. *Australian Journal of Botany* **31**, 383–394.

Stowe, M. K., Tumlinson, J. H., and Heath, R. R. 1987. Chemical mimicry—bolas spiders emit components of moth prey species sex-pheromones. *Science* **236**, 964–967.

Streinzer, M., Ellis, T., Paulus, H. F., and Spaethe, J. 2010. Visual discrimination between two sexually deceptive *Ophrys* species by a bee pollinator. *Arthropod–Plant Interactions* **4**, 141–148.

Streinzer, M., Paulus, H. F., and Spaethe, J. 2009. Floral colour signal increases short-range detectability of a sexually deceptive orchid to its bee pollinator. *Journal of Experimental Biology* **212**, 1365–1370.

Streisfeld, M. A. and Rausher, M. D. 2009. Genetic changes contributing to the parallel evolution of red floral pigmentation among *Ipomoea* species. *New Phytologist* **183**, 751–763.

Sugahara, M., Izutsu, K., Nishimura, Y., and Sakamoto, F. 2013. Oriental orchid (*Cymbidium floribundum*) attracts the Japanese honeybee (*Apis cerana japonica*) with a mixture of 3-hydroxyoctanoic acid and 10-hydroxy-(*E*)-2-decenoic acid. *Zoological Science* **30**, 99–104.

Sugawara, T. 1988. Floral biology of *Heterotropa tamaenesis* (Aristolochiaceae) in Japan. *Plant Species Biology* **3**, 7–12.

Sugiura, N., Goubara, M., Kitamura, K., and Inoue, K. 2002. Bumblebee pollination of *Cypripedium macranthos* var. *rebunense* (Orchidaceae); a possible case of floral mimicry of *Pedicularis schistostegia* (Orobanchaceae). *Plant Systematics and Evolution* **235**, 189–195.

Suinyuy, T. N., Donaldson, J. S., and Johnson, S. D. 2013. Variation in the chemical composition of cone volatiles within the African cycad genus *Encephalartos*. *Phytochemistry* **85**, 82–91.

Sun, H. Q., Cheng, J., Zhang, F. M., Luo, Y. B., and Ge, S. 2009. Reproductive success of non-rewarding *Cypripedium japonicum* benefits from low spatial dispersion pattern and asynchronous flowering. *Annals of Botany* **103**, 1227–1237.

Teck, O. P. 2011. Fly pollination in four Malaysian species of *Bulbophyllum* (Section *Sestochilus*)—*B. lasianthum*, *B. lobbii*, *B. subumbellatum* and *B. virescens*. *Malesian Orchid Journal* **8**, 103–110.

Telles, F. J. and Rodriguez-Girones, M. A. 2015. Insect vision models under scrutiny: what bumblebees (*Bombus terrestris terrestris* L.) can still tell us. *Naturwissenschaften* **102**, 1256.

Thakar, J. D., Kunte, K., Chauhan, A. K., Watve, A. V., and Watve, M. G. 2003. Nectarless flowers: ecological correlates and evolutionary stability. *Oecologia* **136**, 565–570.

Thomas, M. M., Rudall, P. J., Ellis, A. G., Savolainen, V., and Glover, B. J. 2009. Development of a complex floral trait: the pollinator-attracting petal spots of the beetle daisy, *Gorteria diffusa* (Asteraceae). *American Journal of Botany* **96**, 2184–2196.

Thompson, J. N. 1994. *The Coevolutionary Process.* Chicago: Chicago University Press.

Thomson, J. D. 1978. Effects of stand composition on insect visitation in 2-species mixtures of *Hieracium*. *American Midland Naturalist* **100**, 431–440.

Thomson, J. D. 1981. Spatial and temporal components of resource assessment by flower-feeding insects. *Journal of Animal Ecology* **50**, 49–59.

Thomson, J. D. and Plowright, R. C. 1980. Pollen carryover, nectar rewards, and pollinator behavior with special reference to *Diervilla lonicera*. *Oecologia* **46**, 68–74.

Thornhill, R. 1979. Adaptive female-mimicking behavior in a scorpionfly. *Science* **205**, 412–414.

Thornton, R. J. 1807. *New Illustration of the Sexual System of Carolus von Linnaeus and the Temple of Flora, or Garden of Nature*. London.

Tremblay, R. L. and Ackerman, J. D. 2001. Gene flow and effective population size in *Lepanthes* (Orchidaceae): a case for genetic drift. *Biological Journal of the Linnean Society* **72**, 47–62.

Tremblay, R. L. and Ackerman, J. D. 2007. Floral color patterns in a tropical orchid: are they associated with reproductive success? *Plant Species Biology* **22**, 95–105.

Tremblay, R. L., Ackerman, J. D., Zimmerman, J. K., and Calvo, R. N. 2005. Variation in sexual reproduction in orchids and its evolutionary consequences: a spasmodic journey to diversification. *Biological Journal of the Linnean Society* **84**, 1–54.

Turner, J. R. G. 1977. Butterfly mimicry: the genetical evolution of an adaptation. *Evolutionary Biology* **10**, 163–206.

Urru, I., Stensmyr, M. C., and Hansson, B. S. 2011. Pollination by brood-site deception. *Phytochemistry* **72**, 1655–1666.

Vale, A., Navarro, L., Rojas, D., and Alvarez, J. C. 2011. Breeding system and pollination by mimicry of the orchid *Tolumnia guibertiana* in western Cuba. *Plant Species Biology* **26**, 163–173.

van der Niet, T., Hansen, D. M., and Johnson, S. D. 2011. Carrion mimicry in a South African orchid: flowers attract a narrow subset of the fly assemblage on animal carcasses. *Annals of Botany* **107**, 981–992.

van der Pijl, L. and Dodson, C. H. 1966. *Orchid Flowers: Their Pollination and Evolution*. Coral Gables, FL: University of Miami Press.

Van Valen, L. 1973. A new evolutionary law. *Evolutionary Theory* **1**, 1–30.

Vane-Wright, R. I. 1980. On the definition of mimicry. *Biological Journal of the Linnean Society* **13**, 1–6.

Vane-Wright, R. I., Raheem, D. C., Cieslak, A., and Vogler, A. P. 1999. Evolution of the mimetic African swallowtail butterfly *Papilio dardanus*: molecular data confirm relationships with *P-phorcas* and *P-constantinus*. *Biological Journal of the Linnean Society* **66**, 215–229.

Vareschi, E. 1971. Duftunterscheidung bei der Honigbiene—Einzelzell-Ableitungen und Verhaltensreaktionen. *Zeitschrift für Vergleichende Physiologie* **75**, 143–173.

Vereecken, N. J. and Schiestl, F. P. 2008. The evolution of imperfect floral mimicry. *Proceedings of the National Academy of Sciences of the United States of America* **105**, 7484–7488.

Vereecken, N. J. and Schiestl, F. P. 2009. On the roles of colour and scent in a specialized floral mimicry system. *Annals of Botany* **104**, 1077–1084.

Vereecken, N. J., Cozzolino, S., and Schiestl, F. P. 2010. Hybrid floral scent novelty drives pollinator shift in sexually deceptive orchids. *BMC Evolutionary Biology* **10**, 103.

Vereecken, N. J., Mant, J., and Schiestl, F. P. 2007. Population differentiation in female sex pheromone and male preferences in a solitary bee. *Behavioral Ecology and Sociobiology* **61**, 811–821.

Vereecken, N. J., Wilson, C. A., Hoetling, S., Schulz, S., Banketov, S. A., and Mardulyn, P. 2012. Pre-adaptations and the evolution of pollination by sexual deception: Cope's rule of specialization revisited. *Proceedings of the Royal Society B: Biological Sciences* **279**, 4786–4794.

Vincent, B., et al. 2010. The venom composition of the parasitic wasp *Chelonus inanitus* resolved by combined expressed sequence tags analysis and proteomic approach. *BMC Genomics* **11**, 693.

Vogel, S. 1954. *Blütenbiologische Typen als Elemente der Sippengliederug, dargestellt anhand der Flora Südafrikas*. Jena: Fischer.

Vogel, S. 1965. Kesselfallen-Blumen. *Die Umschau der Akademie der Wissenschaften und der Literatur* **10**, 601–763.

Vogel, S. 1972. Pollination von *Orchis papilionacea* L. in den Schwarmbahnen von *Eucera tuberculata* F. *Jahresberichte des Naturwissenschaftlichen Vereins in Wuppertal* **25**, 67–74.

Vogel, S. 1973. Fungus gnat flowers and fungus mimesis. In: *Pollination and Dispersal* (ed. by Brantjes, N. B. M.), pp. 13–18. Nijmegen: Department of Botany, University of Nijmegen.

Vogel, S. 1978. Evolutionary shifts from reward to deception in pollen flowers. *Linnean Society Symposium Series* **6**, 89–96.

Vogel, S. 1978. Pilzmückenblumen als Pilzmimeten [Fungus-gnat flowers mimicking fungi]. *Flora* **167**, 329–398.

Vogel, S. 1993. Betrug bei Pflanzen: Die Täuschblumen. *Abhandlungen der Mathematisch- Naturwissenschaftlichen Klasse* **1**, 5–48.

Vogel, S. 1996. Christian Konrad Sprengel's theory of the flower: the cradle of floral ecology. In: *Floral Biology: Studies on Floral Evolution in Animal-Pollinated Plants* (ed. by Lloyd, D. G. and Barrett, S. C. H.), pp. 44–62. New York: Chapman and Hall.

Vogel, S. and Martens, J. 2000. A survey of the function of the lethal kettle traps of *Arisaema* (Araceae), with records of pollinating fungus gnats from Nepal. *Botanical Journal of the Linnean Society* **133**, 61–100.

von Arx, M., Goyret, J., Davidowitz, G., and Raguso, R. A. 2012. Floral humidity as a reliable sensory cue for profitability assessment by nectar-foraging hawkmoths. *Proceedings of the National Academy of Sciences of the United States of America* **109**, 9471–9476.

von Frisch, K. 1919. Über den Geruchssinn der Bienen und seine blütenbiologische Bedeutung. *Zoologische Jahrbücher Physiologie* **37**, 1–238.

von Hoermann, C., Steiger, S., Mueller, J. K., and Ayasse, M. 2013. Too fresh is unattractive! The attraction of newly emerged *Nicrophorus vespilloides* females to odour bouquets of large cadavers at various stages of decomposition. *PLoS One* **8**(3), e58524.

Vorobyev, M. and Osorio, D. 1998. Receptor noise as a determinant of colour thresholds. *Proceedings of the Royal Society B: Biological Sciences* **265**, 351–358.

Vorobyev, M., Gumbert, A., Kunze, J., Giurfa, M., and Menzel, R. 1997. Flowers through insect eyes. *Israel Journal of Plant Sciences* **45**, 93–101.

Vorobyev, M., Kunze, J., Gumbert, A., Giurfa, M., and Menzel, R. 1997. Flowers through the insect eyes. *Israel Journal of Plant Sciences* **45**, 93–102.

Waddington, K. D. 1981. Factors influencing pollen flow in bumblebee-pollinated *Delphinium virescens*. *Oikos* **37**, 153–159.

Waelti, M. O., Muhlemann, J. K., Widmer, A., and Schiestl, F. P. 2008. Floral odour and reproductive isolation in two species of *Silene*. *Journal of Evolutionary Biology* **21**, 111–121.

Waelti, M. O., Page, P. A., Widmer, A., and Schiestl, F. P. 2009. How to be an attractive male: floral dimorphism and attractiveness to pollinators in a dioecious plant. *BMC Evolutionary Biology* **9**, 109.

Waldbauer, G. P. 1988. Asynchrony between Batesian mimics and their models. *The American Naturalist* **131**, 103–121.

Wall, R. and Fisher, P. 2001. Visual and olfactory cue interaction in resource-location by the blowfly, *Lucilia sericata*. *Physiological Entomology* **26**, 212–218.

Wallis, D. I. 1962. Olfactory stimuli and oviposition in blowfly, *Phormia regina* Meigen. *Journal of Experimental Biology* **39**, 603–&.

Waser, N. M. and Campbell, D. R. 2004. Ecological speciation in flowering plants. In: *Adaptive Speciation* (ed. by Dieckmann, U., Doebeli, M., Metz, M.J., and Tautz, D.), pp. 264–277. Cambridge: Cambridge University Press.

Waterman, R. J. and Bidartondo, M. I. 2008. Deception above, deception below: linking pollination and mycorrhizal biology of orchids. *Journal of Experimental Botany* **59**, 1085–1096.

Weiss, M. R. 1991. Floral colour changes as cues for pollinators. *Nature* **354**, 227–229.

Weiss, M. R. 2001. Vision and learning in some neglected pollinators: beetles, flies, moths, and butterflies. In: *Cognitive Ecology of Pollination* (ed. by Chittka, L. and Thomson, J.D.), pp. 171–190. Cambridge: Cambridge University Press.

Welsford, M. and Johnson, S. D. 2012. Solitary and social bees as pollinators of *Wahlenbergia* (Campanulaceae): single-visit effectiveness, overnight sheltering and responses to flower colour. *Arthropod–Plant Interactions* **6**, 1–14.

Welsford, M. R., Hobbhahn, N., Midgley, J. J., and Johnson, S. D. 2016. Floral trait evolution associated with shifts between insect and wind pollination in the dioecious genus *Leucadendron* (Proteaceae). *Evolution*, **70**, 126–139.

Weston, P., Perkins, A. J., Indsto, J. O., and Clements, M. A. 2014. Phylogeny of Orchidaceae tribe Diurideae and its implications for the evolution of pollination systems. In: *Darwin's Orchids: Then and Now* (ed. by Bernhardt, P. and Meier, R.), pp. 92–154. Chicago: Chicago University Press.

Weston, P. H., Perkins, A. J., Indsto, J. O., and Clements, M. A. 2014. Phylogeny of Orchidaceae tribe Diuridae and its implications for the evolution of pollination systems. In: *Darwin's Orchids: Then and Now* (ed. by Edens-Meier, R. and Bernhardt, P.), pp. 91–154. Chicago: Chicago University Press.

Whitehead, M. R. and Peakall, R. 2013. Short-term but not long-term patch avoidance in an orchid-pollinating solitary wasp. *Behavioral Ecology* **24**, 162–168.

Whitehead, M. R. and Peakall, R. 2014. Pollinator specificity drives strong prepollination reproductive isolation in sympatric sexually deceptive orchids. *Evolution* **68**, 1561–1575.

Whitehead, M. R., Linde, C. C., and Peakall, R. 2015. Pollination by sexual deception promotes outcrossing and mate diversity in self-compatible clonal orchids. *Journal of Evolutionary Biology* **8**, 1526–1541.

Wickler, W. 1965. Mimicry and evolution of animal communication. *Nature* **208**, 519–521.

Wickler, M. 1968. *Mimicry in Plants and Animals*. London: Weidenfeld and Nicolson.

Wiens, D. 1978. Mimicry in plants. *Evolutionary Biology* **11**, 365–403.

Williams, K., Koch, G. W., and Mooney, H. A. 1985. The carbon balance of flowers of *Diplacus aurantiacus* (Scophulariaceae). *Oecologia* **57**, 530–535.

Williamson, G. B. 1982. Plant mimicry: evolutionary constraints. *Biological Journal of the Linnean Society* **18**, 49–58.

Williamson, G. B. and Black, E. M. 1981. Mimicry in hummingbird-pollinated plants? *Ecology* **62**, 494–496.

Willmer, P. 2011. *Pollination and Floral Ecology*. Princeton, NJ: Princeton University Press.

Willson, M. F. 1979. Sexual selection in plants. *The American Naturalist* **113**, 777–790.

Willson, M. F. and Ågren, J. 1989. Differential floral rewards and pollination by deceit in unisexual flowers. *Oikos* **55**, 23–29.

Wong, B. B. M. and Schiestl, F. P. 2002. How an orchid harms its pollinator. *Proceedings of the Royal Society B: Biological Sciences* **269**, 1529–1532.

Wright, G. A. and Smith, B. H. 2004. Variation in complex olfactory stimuli and its influence on odour recognition. *Proceedings of the Royal Society B: Biological Sciences* **271**, 147–152.

Wright, G. A., Skinner, B. D., and Smith, B. H. 2002. Ability of honeybee, *Apis mellifera*, to detect and discriminate odors of varieties of canola (*Brassica rapa* and *Brassica napus*) and snapdragon flowers (*Antirrhinum majus*). *Journal of Chemical Ecology* **28**, 721–740.

Xu, S. and Schlüter, P. M. 2015. Modeling the two-locus architecture of divergent pollinator adaptation: how variation in SAD paralogs affects fitness and evolutionary divergence in sexually deceptive orchids. *Ecology and Evolution* **5**, 493–502.

Xu, S. Q., Schluter, P. M., Grossniklaus, U., and Schiestl, F. P. 2012. The genetic basis of pollinator adaptation in a sexually deceptive orchid. *PLOS Genetics* **8**(8), e1002889.

Xu, S., et al. 2011. Floral isolation is the main reproductive isolation barrier among sexually deceptive orchids. *Evolution* **65**, 2606–2620.

Yeargan, K. V. 1994. Biology of bolas spiders. *Annual Review of Entomology* **39**, 81–99.

Yeates, D. K., Meier, R., and Wiegmann, B. M. 2003. Phylogeny of true flies (Diptera): a 250 million year old success story in terrestrial diversification. *Entomologische Abhandlungen (Dresden)* **61**, 170–172.

Yoshioka, Y., Ohashi, K., Konuma, A., Iwata, H., Ohsawa, R., and Ninomiya, S. 2007. Ability of bumblebees to discriminate differences in the shape of artificial flowers of *Primula sieboldii* (Primulaceae). *Annals of Botany* **99**, 1175–1182.

Young, A. M. 1984. Mechanism of pollination by Phoridae (Diptera) in some *Herrania* species (Sterculiaceae) in Costa Rica. *Proceedings of the Entomological Society of Washington* **85**, 503–518.

Zimmerman, M. 1983. Plant reproduction and optimal foraging: experimental nectar manipulations in *Delphinium nelsonii*. *Oikos* **41**, 57–63.

Zimmerman, M. and Cook, S. 1985. Pollinator foraging, experimental nectar-robbing and plant fitness in *Impatiens capensis*. *American Midland Naturalist* **113**, 84–91.

Zito, P., Sajeva, M., Raspi, A., and Dötterl, S. 2014. Dimethyl disulfide and dimethyl trisulfide: so similar yet so different in evoking biological responses in saprophilous flies. *Chemoecology* **24**, 261–267.

Index